SURFICIAL DEPOSITS
OF THE
UNITED STATES

SURFICIAL DEPOSITS
OF THE
UNITED STATES

CHARLES B. HUNT

VNR VAN NOSTRAND REINHOLD COMPANY
New York

Copyright © 1986 by **Van Nostrand Reinhold Company Inc.**
Library of Congress Catalog Card Number: 85-15692
ISBN: 0-442-23231-4

Manufactured in the United States of America.

Published by Van Nostrand Reinhold Company Inc.
115 Fifth Avenue
New York, New York 10003

Van Nostrand Reinhold Company Limited
Molly Millars Lane
Wokingham, Berkshire RG11 2PY, England

Van Nostrand Reinhold
480 Latrobe Street
Melbourne, Victoria 3000, Australia

Macmillan of Canada
Division of Gage Publishing Limited
164 Commander Boulevard
Agincourt, Ontario MIS 3C7, Canada

15 14 13 12 11 10 9 8 7 6 5 4 3 2 1

Library of Congress Cataloging-in-Publication Data
Hunt, Charles Butler, 1906-
    Surficial deposits of the United States.
    Includes bibliographies and index.
    1. Sediments (Geology)—United States. 2. Geology—United States. I. Title.
QE471.2.H87 1986   551.3   85-15692
ISBN 0-442-23231-4

CONTENTS

FOREWORD

The term "surficial deposits" refers to unconsolidated materials that lie on bedrock or that occur on or near the surface of the Earth. Surficial (superficial) deposits are frequently unstratified and represent the most recent of geologic deposits. Much of the ground surface is occupied by loose or friable deposits of one sort or another, including soil, except in cratonic heartland regions (*e.g.* the Canadian, Baltic, and Angara shields) where these materials were scraped away by continental glaciers during Pleistocene glacial advances. But even these great glaciated shields of exposed bedrock surfaces that characterize so much of the higher latitudes in the Northern Hemisphere retain vestiges of former weathering mantles and unconsolidated deposits. Much of the residuum eroded off the Canadian shield during peaks of glacial activity was, for example, deposited as morainal, fluvioglacial, and loessial materials in the northern tier of the United States. These materials were further modified by surficial (epidiagenetic) processes during interglacial (warm, soil-forming) phases.

This book, providing an introduction to the surficial geology of the United States, is based on a map of surficial deposits prepared by the U.S. Geological Survey for the *National Atlas of the United States of America*. This map of the conterminous United States is, in a very real sense, the *raison d'etre* of this introductory text. Because this full-color map, published at a scale of 1:7,500,000 (one inch equals 120 miles), is carefully explained in this book it should be viewed and referred to while perusing the text.

It is interesting to note that most conventional types of geological maps delineate the nature and extent of bedrock types and structures but often tend to ignore surficial covers. Superficial deposits are, in fact, often a great nuisance to many mapping efforts as interest has traditionally dwelt on the nature of basement rocks. There are good reasons for this focus because the world's mineral resources are predominantly associated with consolidated formations. Awareness of the potential for surficial deposits to contribute to the known mineral base continues to increase as these materials already host about 15% of all mineral production (excluding the mineral fuels). This book offers a new and refreshing look at the nature and composition of the present land surface of the country. Explanations of the geographic distribution of surficial covers also offer greater insight into the more recent geological history of this continental region.

Surficial deposits are, for convenience, commonly divided into three major groups or categories: residual, transitional, and transported deposits. Ancient weathering mantles, often the parent materials of modern soils, form thick clayey deposits in tectonically-geomorphologically stable regions and constitute the greater bulk of the residual deposits. Transitional deposits, those mainly moved under the influence of gravity, include colluvial mantles, landslides,

vii

debris avalanches, and mudflows, among others. Transported deposits include materials moved by the action of water, ice, or wind. The geology of all these types of deposits is complex and may extend back in time to at least the later Tertiary for some of the older accumulations. In spite of the complexities of the surficial geology, this generalized map and introductory text present a useful overview that shows essential relationships of the different kinds of surficial materials.

Although intended for use by students and laymen, many professionals as well will find this volume useful in their work, especially when understanding of the "big picture" is required. Many surficial deposits themselves are composed of ancient weathering mantles that contain paleosols which in turn may constitute the initial materials of contemporary pedogenesis. Although there is a clear nexus between surficial materials and soils, the nature of the deposits alone is explicit on the map and in the accompanying text.

CHARLES W. FINKL, JNR.

PREFACE

This book is the outgrowth of an effort to prepare an explanation for the 1/7,500,000 overview map of surficial deposits of the United States which was prepared for the U.S. Geological Survey's National Map Atlas. The scale indicates that 1 inch on the map equals 7,500,000 inches on the ground, or about 120 miles. Because the scale is small, the map is of only limited use to specialists, but because the map information is technical, the nonspecialist needs an explanation that is more complete than can appear on the map. Accordingly this book is an elementary text, a primer, addressed to nonspecialists and students. It is intended to be an introduction to the subject.

Surficial deposits, the mantle of loose material on the surface and covering most bedrock, are of major importance in many kinds of engineering projects, in agriculture and forestry, and in all kinds of land use planning and development. Although detailed maps are needed for the land planning, an understanding of the subject should help planners determine when particular kinds of ground are important in connection with a planned use. The importance of land planning and the importance of surficial deposits in that planning was demonstrated dramatically by the terrain intelligence folios prepared for the Corps of Engineers by the Military Geology Unit of the Geological Survey during World War II. Those folios consisted of quadrangle maps with explanations summarizing how the ground conditions affect land use. While the emphasis was on combat engineering, the methodology pointed the way for applying geology to all kinds of land use planning.

In this book surficial deposits are divided into four major categories: 1) untransported deposits, those formed more or less in place by weathering (residuum) or by organic activity (as in marshes and swamps); 2) transitional deposits, those displaced by gravity (such as colluvium, landslides, debris avalanches, or mudflows) (The map, unfortunately, includes these with untransported deposits, although the deposits have moved.); 3) transported deposits, which include shore, glacial, stream, lake, and wind deposits; and 4) miscellaneous kinds of deposits including basalt, clinker due to the burning of coal beds, hot spring deposits, and the small patches of bedrock shown on the overview map.

CHARLES B. HUNT

SURFICIAL DEPOSITS AND THEIR IMPACT ON LAND USE

Surficial deposits constitute the mantle of unconsolidated material covering most of the continent and overlying the bedrock forming the structural plate that is our continent. Surficial deposits are the equivalent of what engineers refer to as soil, that is, surface materials that can be excavated without blasting. The term *soil* is ambiguous, even in the usage of the Department of Agriculture and agricultural colleges, where *soil* may refer only to those agriculturally important top layers of a surficial deposit, the subject of soil surveys, or may refer to all parts of a surficial deposit subject to erosion.

This book is organized around the groups of surficial deposits shown on the U.S. Geological Survey's 1/7,500,000 map, a copy of which accompanies the book. A map at that scale (almost 120 miles [193 km] to the inch) is necessarily too generalized to be of much use to specialists, yet the subject matter requires technical explanation far beyond what can be included in a map explanation. Accordingly, this book, an introduction to surficial deposits, is intended to provide an overview commensurate with the scale of the map. The deposits are described in four categories: untransported, transitional, transported, and miscellaneous deposits (Fig. 1). Their distribution among the country's physiographic provinces is shown in Fig. 2.

Untransported deposits (Chapters 1 and 2) are further divided into four kinds: thick residual deposits over deeply weathered bedrock, found mostly in the warm and humid southeastern United States; thin residuum over little weathered bedrock in the arid and semiarid western states; organic deposits, especially those in swamps and marshes; and some other important deposits, such as caliche and desert pavement, that could not be shown on the map.

Thick residual deposits in the southeastern United States (Chapter 1), referred to as saprolite, are the product of deep weathering of the bedrock; saprolite tends to be clayey, which is the common product of the decomposition of most minerals. As a consequence, the ground is subject to severe erosion

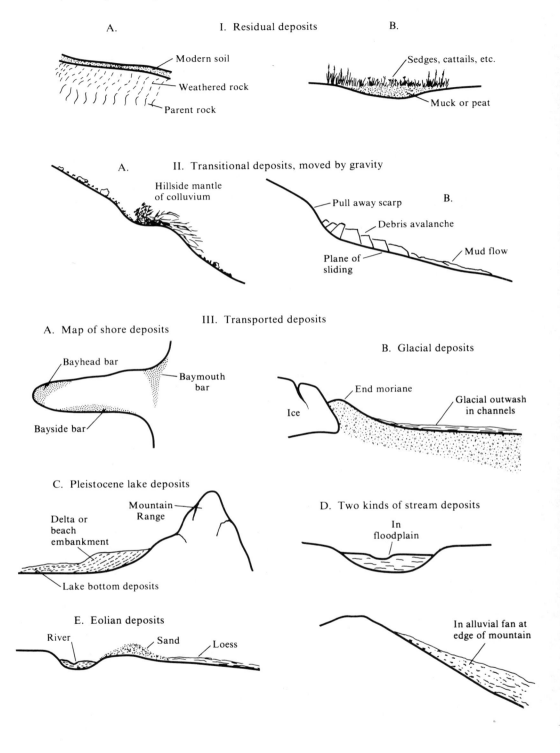

A.   I. Residual deposits   B.

Modern soil

Weathered rock

Parent rock

Sedges, cattails, etc.

Muck or peat

A.   II. Transitional deposits, moved by gravity

Hillside mantle of colluvium   B.

Pull away scarp

Debris avalanche

Mud flow

Plane of sliding

III. Transported deposits

A. Map of shore deposits

Bayhead bar

Baymouth bar

Bayside bar

B. Glacial deposits

Ice

End moriane

Glacial outwash in channels

C. Pleistocene lake deposits

Mountain Range

Delta or beach embankment

Lake bottom deposits

D. Two kinds of stream deposits

In floodplain

E. Eolian deposits

River   Sand   Loess

In alluvial fan at edge of mountain

Figure 1. Diagrammatic sections (except III*A*) illustrating the principal kinds of surficial deposits. Surficial deposits divide into three major kinds: I, residual deposits; II, transitional deposits; and III, transported deposits.

I. Residual deposits.

I*A*. Thick, generally clayey residuum characterizes the warm, humid, mid-Atlantic and southeastern states. Water containing organic acids enters the ground and alters the minerals to clay. These are ancient soils and are the parent material for modern soils. In the arid and semiarid western states, weathering is shallow, residuum is thin, and the parent rock minerals are only slightly altered.

I*B*. Where there are topographic sags, water collects to form marshes or swamps, and the organic matter accumulates as muck or peat.

II. Transitional deposits move chiefly by gravity (included with "untransported deposits" on the map).

II*A*. Colluvium is the mantle of loose material working down hillsides by creep, which is caused by washing, wetting and drying, frost or other ground heaves, falling trees, animal burrows, animal tracks, and so on.

II*B*. Large masses of colluvium may pull away and move as a debris avalanche. Water escaping from the avalanche can create a mudflow. Large masses of bedrock may become detached and move down as landslides.

III. Transported deposits may have been moved by shore or lake currents, glacial ice, streams, or wind.

III*A*. Map shows positions of baymouth, bayhead, and bayside bars.

III*B*. Cross section showing the front of a glacier that has pushed up an end moraine. Beyond the front is glacial outwash in stream channels. Under the outwash may be ground moraine deposited by the ice as it melted back to the position shown.

III*C*. Lake deposits, especially those of the Pleistocene lakes, include deltas, beach embankments, spits, bars, or other shore deposits. The lake bottom sediments are fine-grained silt and clay and may include evaporites.

III*D*. Stream deposits may be in floodplains or in alluvial fans at the base of mountains, including the Pliocene and older stream deposits on the Great Plains (osg on the map).

III*E*. Wind-transported deposits form sand dunes or, if the material is fine grained, loess.

wherever it is disturbed, as in plowed fields in rural areas and at building sites in urban areas. At most construction sites in the eastern United States, for example, mud is tracked onto the adjoining highway or street. Saprolite forms productive farmland, but these lands require fertilizing because they are so weathered and so many nutrients have been depleted.

Saprolite is commonly many feet thick and may be as deep as 100 feet (33m). The several mineralogical and textural varieties of saprolite shown on the map reflect various kinds of bedrock that were weathered.

Saprolite forms our oldest surficial deposits. All saprolite antedates the last Pleistocene glaciation—that is, it is pre-Wisconsinan (see Table 1). Much of it is pre-Pleistocene. Saprolite is rarely found in those northern States that were glaciated because any that had formed was eroded by the ice and incorporated into the glacial deposits. However, isolated pockets of saprolite are occasionally found in excavations that extend below the cover of glacial or other deposits.

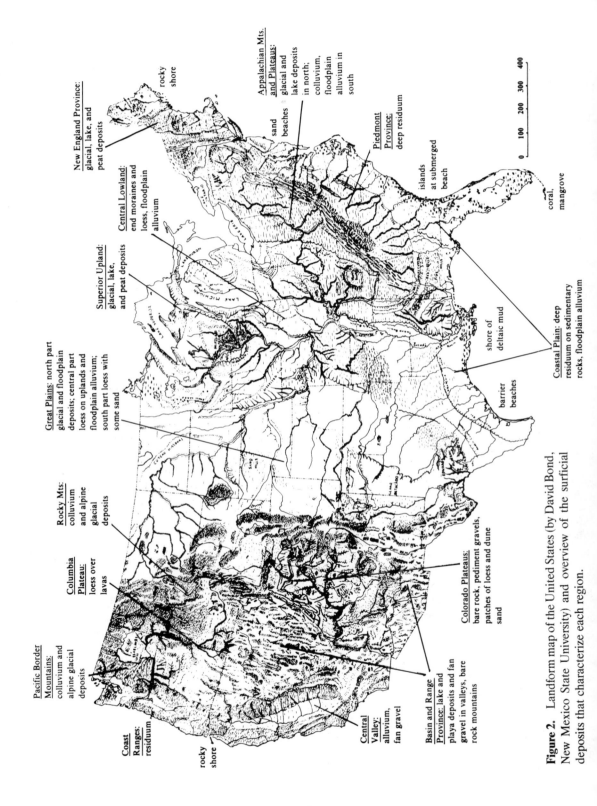

Figure 2. Landform map of the United States (by David Bond, New Mexico State University) and overview of the surficial deposits that characterize each region.

Pacific Border Mountains: colluvium and alpine glacial deposits

Coast Ranges: residuum

rocky shore

Columbia Plateau: loess over lavas

Rocky Mts: colluvium and alpine glacial deposits

Great Plains: north part glacial and floodplain deposits; central part loess on uplands and floodplain alluvium; south part loess with some sand

Central Valley: alluvium, fan gravel

Basin and Range Province: lake and playa deposits and fan gravel in valleys, bare rock mountains

Colorado Plateaus: bare rock, pediment gravels, patches of loess and dune sand

New England Province: glacial, lake, and peat deposits

rocky shore

Superior Upland: glacial, lake, and peat deposits

Central Lowland: end moraines and loess, floodplain alluvium

Appalachian Mts. and Plateaus: glacial and lake deposits in north; colluvium, floodplain alluvium in south

sand beaches

Piedmont Province: deep residuum

islands at submerged beach

coral, mangrove

shore of deltaic mud

barrier beaches

Coastal Plain: deep residuum on sedimentary rocks, floodplain alluvium

0 100 200 300 400

4

Table 1.   Geologic Timetable of the Pleistocene Glaciations and Interglaciations and of the Earlier Cenozoic in the United States

		Epoch	Glaciation	Interglaciation	Began: Estimated Age, B.P. (before present)
Cenozoic Era	Quaternary Period	Holocene			11,000
			Wisconsinan		70,000
		Pleistocene		Sangamon	
			Illinoian		
				Yarmouth	
			Kansan		500,000
				Aftonian	750,000
			Nebraskan		1,000,000
				Blancan	1,600,000
			– – –	– – –	2,000,000
	Tertiary Period	Pliocene			5,300,000
		Miocene			23,700,000
		Oligocene			36,600,000
		Eocene			57,800,000
		Paleocene			66,400,000

The second kind of residuum, also described in Chapter 1, is found in the arid and semiarid western states where weathering of the bedrock has not been deep. This situation partly reflects the comparative scarcity of moisture sufficient to penetrate deeply into the bedrock, but it also reflects the comparative youthfulness of the western terrain. The textures and mineralogy of these western deposits closely reflect those of the parent bedrock, which may be any kind of igneous, sedimentary, or metamorphic rock. In the western United States the kind of bedrock is immediately apparent from the residuum on it; in the southeastern United States a much closer look is needed. The thin residuum is discontinuous but is very important for developing and holding plant cover. Augmented with atmospheric dust, it forms the medium in which roots can spread and obtain the nutrients needed for plant growth.

Thin residuum, like saprolite, is easily eroded, and generally patches of the stripped bedrock surface are exposed. Nevertheless, the extent of the preserved residuum, because of its vegetative cover, is an important factor in determining the capacity of the ground for grazing, farming, or other use.

A third kind of untransported deposit is of organic origin (Chapter 2). These are young deposits, most of them still forming today in marshes and ponds. Such poorly drained ground is extensive in the coastal plain in the southeastern United States and in depressions in the hummocky topography that characterizes the glacial deposits in the northern states. Some of the deposits, north and south, are sources of peat (black triangles on the map). These deposits pose foundation problems where they must be crossed by highways. Perhaps the greatest importance of these wetlands is in providing favorable habitats for wildlife, including mosquitoes.

Other important materials of organic origin are found in a belt along the south Atlantic and Gulf coasts where strong acid leaching of the ground has removed organic matter from the surface and redeposited it as a thin layer of humus at the top of the shallow water table. Such subsurface organic deposits are referred to as groundwater Podzols (gp on the map). Because the surface layers are strongly leached, the ground requires considerable fertilizer to be useful for agriculture. Furthermore, the ground is poorly drained.

Other kinds of residual deposits are important locally but are not shown on the map because of the scale. The most extensive of these is caliche, calcium carbonate cementing near-surface layers of gravel, sand, and silt. In the southwestern United States the deposit may be thick enough and tough enough to be quarried for road metal. Generally, because of the cement, the layer forms a cap that resists erosion. It is a conspicuous layer in all kinds of surficial deposits throughout the southwestern United States. Where the caliche is tight, it can increase surface runoff by obstructing seepage of water into the ground.

Another extensive kind of deposit not shown on the map, known as desert pavement, is also a feature of southwestern United States. It is especially well developed and extensive on the gravel fans of the Basin and Range Province. Desert pavement is a lag deposit of closely set stones. The stones become pushed to the surface of the gravel fan by frost and other heaving of the ground and by the removal of fine sediments by surface runoff or by wind. In places the surface is smooth enough to form a true pavement that accelerates runoff so greatly the ground is bare pavement without vegetation. The smooth pavements are generally old enough to be blackened with a stain of iron and manganese oxide known as rock stain or desert varnish, also described in Chapter 2.

Other residual deposits not shown on the map but briefly described in Chapter 2 include gossans over ore deposits, turquoise, patterned ground attributable to intensive frost action at alpine summits, travertine (calcium carbonate) deposits that occur in mounds or layers at springs, cement in beach deposits of the Pleistocene lakes in the Great Basin, and cave deposits.

Transitional between the untransported and transported deposits are those that have moved down slopes by gravity (Chapter 3). The only category of these shown on the map is colluvium (co on the map; subscripts indicate the kind of parent rock); the colluvium has been moved by gravity and is not properly an untransported deposit although so shown on the map, nor is it a truly transported one. The mechanism by which colluvium moves downslope, known as mass wasting, involves slow downhill creep of stones and other debris as a result of freezing and thawing, wetting and drying, and surface washing. Any disturbance of the sloping ground caused by man, animals, or falling trees dislodges stones and earth and causes them to move downhill. Frost lifts stones, however slightly, and when they settle they move down the slope. Trains of debris from layers of gravel extend downslope from their source. Colluvium is unstable ground. Some colluvial slopes are suitable for orchards, but most such ground is best left as forest or woodland. Such hillsides, where grazed, quickly develop terracettes. The angle of repose of loose, dry, mineral grains, like sand, is a little more than 30°. More earthy material held together by roots is likely to have an angle of repose of less than 20°.

Related to colluvium but not shown on the map because of their limited size are landslides, debris avalanches, mudflows, boulder fields, and collapse structures in limestone. Some of these are very important geological hazards. They are briefly described in Chapter 3.

Transported deposits (Chapters 4 to 8) are those formed by materials that were weathered or eroded at one place and deposited elsewhere after transportation by coastal shore currents, glacial ice, streams, lake currents, or wind. Man's activities commonly cause accelerated erosion by modifying shore currents, stream flow, surface runoff, or exposure to wind (as by plowing or otherwise clearing land). Each transporting medium produces distinctive kinds of deposits, and the principal occurrences of the five kinds are shown on the map.

Sandy shores (Chapter 4) are indicated on the map by a hachured line. They characterize the Gulf Coast and the Atlantic Coast as far north as Cape Cod. The New England coast north of Cape Cod and most of the Pacific Coast is characterized by rocky headlands separating short stretches of sandy beach. The sand beaches, of course, are favorite vacation lands. Along the Atlantic and Gulf coasts they are subject to severe flooding and washing when hurricanes strike the coast, especially those storms that coincide with high tides. Similar flooding occurs along the Pacific Coast when there are severe storms or tsunamis (the large water waves generated by earthquakes).

Associated with the beaches are backshore deposits—wetlands (bm on the map)—that are important for wildlife. They are becoming a major political issue because population pressure is bringing increased demand for land development. The wetlands are unattractively buggy, but they play a major role in the nation's ecology because of the habitat they provide for wildlife, particularly migratory waterfowl. Both the wetlands and the sand beaches with which they are associated are young (Holocene) deposits (Table 1). They have formed since the sea level rose to its present level following the lowered sea level that accompanied the Pleistocene glaciations.

The contrast in surficial deposits along the eastern and western coasts of the United States reflects the westward drift of the continent. The rocky headlands of the Pacific Coast and the narrow continental shelf off that shore constitute the leading edge of the westward drifting continent; the gently sloping coastal plain along the Atlantic and Gulf coasts and the broad, submerged continental shelf offshore from them constitute the trailing edge. The rocky coast northeast of Cape Cod is attributable to northeastward tilting of that part of the continent and the submergence of the coastal plain as well as the continental shelf north of the cape.

Pleistocene glacial deposits (Chapter 5), which blanket the northern United States, generally are stony and supplied the many miles of stone fences in New England. In the Middle West the ground is less stony, and loess-covered (Chapter 8) till plains there provide rich agricultural ground, our corn belt. Glacial deposits also occur near the tops of many western mountains. The map also shows some of the existing snowfields on the high mountains in the western United States (black square symbols).

Stream deposits (Chapter 6) that can be shown on the map are of three principal kinds: floodplains along rivers (al), alluvial fan deposits (fs and fg) sloping from the foot of desert mountains, and Pliocene and older stream deposits on the Great Plains (osg). The latter form uplands traversed by floodplain deposits of the younger rivers flowing eastward to the Mississippi River.

Floodplain deposits, as the name implies, are subject to flooding. They are rich agricultural ground because the ground has plentiful nutrients for plant growth and plenty of water. In urban areas such land is best used for parks and playgrounds, which are increasingly needed as recreational areas, and such use provides a means for temporary storing of floodwaters to relieve flooding downstream. Alluvial fan deposits are well drained and dry. On most fans in the western United States water is too scarce and the ground too stony to be used for farming. The fans are mostly range land for grazing by livestock.

Most of the lake deposits (l) shown on the map and described in Chapter 7 are those of the great Pleistocene lakes in the Great Basin, but the organic deposits in the glaciated part of the northern United States are also in part lacustrine. The backshore deposits along the coasts are somewhat like lake deposits. The Pleistocene lake deposits in the Great Basin, especially the deposits of Lake Bonneville in western Utah and of Lake Lahontan in western Nevada, include extensive areas of evaporites, which are sources of various salts. Surrounding the evaporites are lake-bottom silts, which are good farmland. On the higher parts of the alluvial fans and along the beaches on the hillsides bordering the old lakes are sand and gravel deposits forming embankments, bars, and spits that were built by the lakeshore currents. These are major sources of sand and gravel.

Sand and gravel deposits, whether formed by streams or along lake shores, constitute one of the most valuable—dollar-wise—resources of the country. The value of sand and gravel produced in a typical year—1978, for example—is about two-thirds the value of Portland cement production, 15 percent of the value of coal produced, about equal to the value of copper produced before the recession, and many times the value of our gold and silver production. Yet by far the most valuable and productive surficial deposits are the wind-deposited silts (wl and es on the map) known as loess (Chapter 8).

When considering wind-deposited materials one is likely to think first of sand dunes, but the most extensive and most important deposits form the richly productive farmland in the central United States where the major part of America's agriculture is based. The value of America's farm production is immense, and most of that produce is from loess in the Middle West.

Other important loessial deposits also are richly productive, such as the loess on the Columbia Plateau and the much smaller belt on the Colorado Plateau in southeastern Utah and southwestern Colorado. The Colorado Plateau deposits, a major producer of pinto beans, were derived from deserts to the southwest, and in having a desert source, this loess resembles that in the eastern Mediterranean countries. Loess in the central United States and on the Columbia Plateau, on the other hand, was derived from beds of streams that discharged glacial meltwaters. When floods ebbed during the dry seasons, those streambeds became barren mud flats and the westerly winds could transport silts eastward from the streambeds. Eastward from each river in the central United States, the loess becomes thinner and fine grained. Along the Mississippi River the occurrence of such loess east of the river valley shows even on the 1/7,500,000 scale map (es).

Although the loess is rich farmland, it is easily eroded and is the source of much of the mud that characterizes our middle western rivers. Mark Twain wrote (in *Life on the Mississippi*), ". . . every tumblerful of it holds nearly an acre of land . . . . If you will let your glass stand half an hour, you can separate the land from the water easy as Genesis . . . ."

In addition to the loess, there are deposits of sand (s, osg) that locally form dunes. Such sand is extensive along our shores, as already noted. It is extensive in the Texas Panhandle and in southeastern New Mexico. At White Sands, New Mexico, the deposits are unusual in being composed of gypsum grains rather than the more usual quartz sand. Similar gypsum dunes occur along the east side of the Great Salt Lake desert, although that belt of deposits is too narrow to be shown on the map.

The miscellaneous deposits (Chapter 9) include basaltic lavas (b). Those that are of the middle or late Tertiary age generally have surfaces that have been smoothed by weathering and by deposition of silt by wind or by surface washing. Young lavas of late Pleistocene or Holocene age tend to be rough with jagged rock surfaces. The map does bring out the fact that the most easterly of the obviously volcanic rocks in the United States are in northeastern New Mexico.

The map shows very little bedrock except in the desert mountains of the Basin and Range Province. This is because our mountains, even including the supposedly rocky ones, are largely mantled by gravity deposits, mostly colluvium, and outcrops of the bedrock are too small to be shown. Only about 5 percent of the Appalachians, including the Blue Ridge and Appalachian plateaus, is bare rock. The dense deciduous forests in those areas are rooted in colluvium. The same is true in the mountains of the western United States where land inventories indicate about 15 percent of bedrock. The map shows that bedrock (br) is extensive in the canyon country of the Colorado Plateau.

Hot-spring deposits (o) are numerous, especially in the western United States, where earth movements have been recent enough for faults to provide deep water circulation to buried, heated rocks, or where there still remain shallow hot rocks, as at Yellowstone. At a few places the heated water is developed as geothermal energy.

On the northern Great Plains, lignite and subbituminous coal beds have ignited spontaneously, and their burning has baked the overlying beds to clinker (kk), quite like the clinker coming from a coal furnace. The deposits are widespread, although individual ones are likely to be small. On the map they can be indicated only diagrammatically, to record the presence of such deposits in the region.

Surficial deposits record not only the geologic history of the Cenozoic, especially the last couple of million years (Table 1), but also involve every process—physical, chemical, and biological—operating on and in the earth. If the present is the key to past, as has long been a geologic dictum, knowledge of surficial deposits and the processes they record is essential for understanding the more ancient geologic record of bedrock.

Surficial deposits have immediate practical importance in at least eight other very different ways: economic geology, water supply, waste disposal, foundation problems, soils and weathering, sedimentation and erosion, marine science, and general environmental and engineering geology.

Economic geology, as that subject is taught in most geology departments, is thought of as a *hardrock* subject, but according to the U.S. Bureau of Mines, up to 15 percent of our mineral production (other than mineral fuels) is from surficial deposits. Most of this production is sand, gravel, and other construction materials, from glacial, stream, or lake deposits, but other mineral resources like placers, residual ores (e.g. bauxite, phosphate), and evaporites are also important. Weathering is a factor in evaluating mineral deposits, for secondarily enriched deposits are to a considerable degree a legacy of Cenozoic weathering. The downward succession of oxidized, secondarily enriched, and unaltered primary ores at porphyry copper pits in the southwestern United States is actually a deep weathering profile in mineralized rock. The residuum (I.A) shown in Figure 1 could be an example (cf. Figs. 1.1A and 2-9).

The importance of surficial deposits in prospecting for mineral deposits has received increased attention because of interest in geochemical prospecting. This involves a search for buried mineral deposits by determining the presence of trace elements in surficial deposits, in accumulator plants growing on the deposits, and in waters seeping through them.

Our water requirements and problems of waste disposal are prodigious. Matters of water supply and sewage disposal hinge importantly on surficial deposits. O. E. Meinzer, widely regarded as the founder of groundwater hydrology in this country, once wrote (1923, p. 222): ". . . it would probably not be an exaggeration to say that [the collection of surficial, mostly Quaternary deposits] is as important as all other systems taken together. It lies at the surface throughout the largest area, supplies the most wells, and affords the greatest quantities of water."

Foundation problems for heavy construction, whether highways, bridges, airstrips, buildings, dams, or other structures, involve the thickness, lithology, drainage capabilities, and engineering properties of the surficial deposits. Ground stability of surficial deposits, their susceptibility to landsliding, avalanching, settling by compaction, heaving by swelling of clays, and creep by gravity sliding are common problems in engineering, and, as will be brought out, the problems are not confined to steep slopes.

Sedimentation and erosion problems arise in every kind of environment— beaches, harbors, lakes, reservoirs, rivers, hillsides, river floodplains, alluvial fans, and dust bowls. The gullying and arroyo cutting that has occurred widely throughout the country in recent decades has been blamed largely on human use of the land, but Quaternary geologic history records numerous prehistoric episodes of similar erosion. To maintain perspective in developing sensible controls against land abuse, the role of natural processes needs to be understood. Damage by erosion will not be lessened by oversimplified assumptions about causes and cures.

Surficial deposits are a major factor in plant geography, one of the biologic aspects of geology. While climate is of first order importance in plant distribution, the kind of ground—that is, the surficial deposits (especially their water properties) —determines the distribution of plants within a climatic region. Some plants (phreatophytes) grow only where the roots can reach the water table. And, if there is a difference in the quantity and kind of salts in the water, these differences are faithfully reflected by differences in the plant species because of differences in their salt tolerances. Other plants (xerophytes) depend on the ephemeral water in the vadose zone above the water table and must be capable of surviving protracted periods of drought. Xerophytic species reflect differences in the supply and retention of ephemeral moisture and are zoned with respect to those differences because different species have different capabilities of resisting drought. Both phreatophytic and xerophytic plants, especially the woody perennials, are excellent indicators of ground conditions and assist in indicating the extent of particular surficial deposits.

The chronology and other stratigraphy of surficial deposits is based on the principle of superposition of strata—younger deposits overlie older ones—and on correlations based on the lateral tracing of individual deposits from one locality to another. Gravel or sand may be deposited in one place while muds are being deposited elsewhere; when traced laterally the different kinds of deposits grade into one another. Such contemporaneous differences in deposits are termed facies. In brief, surficial deposits are divided into mappable, lithologic units the same way that bedrock formations are divided.

Fossils in a series of formations record the sequence of species. The paleogeographic conditions, including the paleoclimates, are recorded by those fossil fauna and flora and by the physical and mineralogical features of the deposits.

Time units in stratigraphy are defined in several ways: by the first appearance of a species considered diagnostic of the younger formations; by the last appearance of a species considered diagnostic of the earlier formations; or on the basis of maximum change in the fauna or flora. The Cenozoic (Table 1) is commonly taken to begin at the disappearance of the big reptiles, the dinosaurs. Other time units are defined on the basis of the greatest change in fauna. The Pleistocene-Holocene boundary selected here is placed at the horizon of the greatest change in fauna; it emphasizes faunal change rather than the appearance or disappearance of a particular diagnostic species. Elephant, camel, horse, long-horned bison, and other animals disappeared from North America at the end of the Pleistocene. In terms of the geologic record they disappeared abruptly. Why they became extinct is not known. Some would blame humans for the extinctions, although humans may have contributed to the extinctions, similar extincti,ons are recorded repeatedly in the geologic past.

Archaeological remains are useful for dating Holocene formations. For example, in most of the United States pottery was not introduced until after about 1 A.D. A surficial deposit that contains pottery is thus a Late Holocene sediment.

The following chapters summarize the principal features shown on the 1/7,500,000 map of the United States plus some examples of the kind of details that may be expected within the broad units shown on the map.

REFERENCES CITED AND ADDITIONAL BIBLIOGRAPHY

Cargo, D. N., and Mallory, B. F., 1974, *Man and His Geologic Environment,* Addison-Wesley Publishing Co., Reading, Mass. 548p.

Charlesworth, J. K., 1966, *The Quaternary Era* (2 vols.), St. Martins Press, New York, 1700p.

Costa, J. E., and V. R. Baker, 1981, *Surficial Geology,* John Wiley and Sons, New York, 498p.

Cooke, R. U., and J. C. Doornkamp, 1974, *Geomorphology in Environmental Management*, Clarendon Press, Oxford, 413p.

Dawson, A. D., 1982, *Land-use Planning and the Law*, Garland STPM Press, New York and London, 246p.

Doering, J. A., 1960, Quaternary surface formation of southern part of Atlantic Coastal Plain, *Jour. Geology* **68:**182-202.

Flawn, P. T., 1970, *Environmental Geology*, Harper and Row, New York, 313p.

Flint, R. F., 1957, *Glacial and Pleistocene Geology,* John Wiley and Sons, New York, 553p.

Hay, O. P., 1923, 1924, 1927, The Pleistocene of North America and Its Vertebrated Animals (3 vols.), *Carnegie Inst. Washington Pub. 322, 322A, and 322B.*

Hunt, C. B., 1972, *Geology of Soils,* W. H. Freeman and Co., New York, 344p.

Jenny, H., 1941, *Factors of soil formation*, McGraw-Hill, New York, 281p.

Legget, R. F., 1962, *Geology and Engineering,* McGraw-Hill, N. Y. 884p.

Mabbutt, J. A. 1977, *Desert Landforms*, MIT Press, Cambridge, Mass., 340p.

Meinzer, O. E., 1923, The occurrence of ground water in the United States, *U.S. Geol. Survey Water-Supply Paper 489,* 321p.

Millot, G., 1970, *Geology of clays*, Springer-Verlag, New York, 429p.

Morrison, R. B. and others, 1957, In behalf of the Recent, *Am. Jour. Sci.* **255:**385-393.

Smith, R. T., and K. Atkinson, 1975, *Techniques in Pedology,* Paul Elek Ltd., London, 213.

Terzaghi, K. and R. B. Peck, 1967, *Soil Mechanics in Engineering Practice,* John Wiley, New York, 729p.

United Nations Food and Agriculture Organization, 1968, Definitions of Soil Units for the Soil Map of the World, UNESCO, 73p.

U.S. Dept. of Agriculture, 1938, Soils and Men, Washington, D.C., 1232p. U.S. Dept. of Agriculture, 1951, *Soil Survey Manual*, Washington, D.C., 503p.

Wright, H. E., Jr., and D. G. Frey, eds., 1965, *The Quaternary of the United States,* Princeton University Press, Princeton, N.J., 922p.

Wright, H. E., ed., 1983, *Late Quaternary Environments of the United States*, University of Minnesota Press, Minneapolis, Minn.

Yaalon, D. H., ed., 1971, *Paleopedology — Origin, Nature and Dating of Paleosols* Israel Universities Press, Jerusalem, 350p.

Note: About weathering, see, for example, *Soil Science Society of America Journal,* Jan. and Feb., 1985, 284p.

PART

I

UNTRANSPORTED DEPOSITS

RESIDUUM: THICK AND THIN

SAPROLITE

Saprolite is ancient residual soil and weathered rock formed by alteration of rock materials to clays and other residual minerals. The clayey products, largely kaolin in the United States, are derived from weathering of such minerals as feldspars, mica, amphibole, and pyroxena; if much quartz was present in the parent rock, the saprolite is sandy. A curious and distinctive feature of saprolite is that the arrangement of the alteration products preserves the original rock structure.

Saprolite is widespread and most conspicuous on the Piedmont Plateau, between the Blue Ridge and the coastal plain (Fig. 2), where it has developed on various kinds of metamorphosed Paleozoic and Precambrian rocks. It is equally widespread on the coastal plain but is less noticeable there because the sedimentary formations on which it developed are themselves in large part derived from the erosion of older saprolitic formations. Saprolite is patchily preserved in the Appalachian Mountains, especially over limestone formations on the flat floors of the valleys of the Valley and Ridge Province between the Blue Ridge and Allegheny plateau. The scarcity of saprolite on steep hillsides probably reflects relatively vigorous erosion on the slopes. Remnants of saprolite on the Cumberland Plateau are numerous and extensive, notably in the southern part.

Saprolite is pre-Wisconsinan in age (see Table 1) and ends northward at the edge of the Wisconsinan drift, where it is overlapped by the drift — unequivocal evidence that it is older than the Wisconsinan glaciation. North of the drift border, saprolite has been found at a few locations where erosion or excavations have revealed isolated pockets of saprolite that escaped erosion by the ice and were preserved beneath its deposits.

The 1/7,500,000 map shows that saprolite also is extensive in the central United States south of the glacial deposits, especially on limestone formations in southern Indiana, western Kentucky, and on the Ozark Plateau. Such saprolite on limestone is equivalent to the European terra rossa. Still another area of saprolite shown on the map extends from north-central and northwestern California northward along the Coast Ranges of Oregon and Washington.

In semiarid and arid parts of the western United States, pre-Wisconsinan weathering has also involved clayey alteration of hard-rock parent materials, but these deposits are associated with substantial accumulations of calcium carbonate (caliche), supplied in part by wind-deposited calcareous dust.

Microscopic studies of fresh rock, weathered rock, and saprolite reveal the mineralogic changes caused by the weathering. Potash feldspars become clouded with clay (kaolinite), while biotite may remain fresh. Soda-calcium feldspars become clouded with kaolinite while hornblende and augite may remain fresh. Blue-green hornblende and epidote survive where saprolite is developed over amphibolite and gabbro, but virtually all pyroxene has disappeared. Garnet is scarce in most samples although abundant in most of the parent rocks.

The term *saprolite* refers especially to the thoroughly decomposed, earthy, untransported residuum on the crystalline rocks of the southern part of the Piedmont Plateau, but the term has been extended to other areas where hard rocks have weathered to clay. From Pennsylvania to Alabama, fields are red and yellow, and road cuts expose red and yellow saprolite. It is generally tens of feet thick, and locally may be more than 100 feet (33m) thick. The average thickness increases southward. In the Baltimore-Washington area it averages 25 to 50 feet (7.6 to 15.2 m) thick; in the Carolinas and Georgia depths greater than 100 feet (33 m) are not uncommon. Whether the greater thickness southward is due to deeper weathering or to less erosion is not known.

Where fully developed, saprolite has orderly layers (Fig. 1-.1A). At the bottom is fresh, undecomposed bedrock. Above this is a variable thickness of weathered bedrock. This layer is tough and must be broken by a hammer, but the rock is sufficiently weathered that, when struck by a hammer, it does not

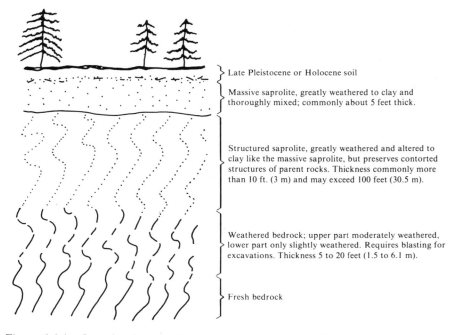

Late Pleistocene or Holocene soil

Massive saprolite, greatly weathered to clay and thoroughly mixed; commonly about 5 feet thick.

Structured saprolite, greatly weathered and altered to clay like the massive saprolite, but preserves contorted structures of parent rocks. Thickness commonly more than 10 ft. (3 m) and may exceed 100 feet (30.5 m).

Weathered bedrock; upper part moderately weathered, lower part only slightly weathered. Requires blasting for excavations. Thickness 5 to 20 feet (1.5 to 6.1 m).

Fresh bedrock

Figure 1-1A. Layering in saprolite usually includes an oxidized zone of weathered bedrock at the base overlain by clayey saprolite that preserves the structures of the parent rock. Above this is a massive, clayey saprolite, and at the surface is 15-25 inches (38-63 cm) of Holocene or Late Pleistocene soil. The surface layers constitute the agriculturists' "soil."

ring like fresh rock but gives a dull thud. This weathered rock is discolored brown or yellow with hydrated iron oxides, especially along partings or cracks (joints). Altered feldspars are still firm. The density of the weathered rock is 5 to 10 percent less than that of the fresh rock.

In dense rocks this layer of weathered bedrock is thin, but in porous rock it may be many feet thick. Where the bedrock is a coarse grained acidic rock (high in silica, like granite), the weathered layers are sandy. Individual minerals may look fresh, yet the rock disintegrates into particles, probably because of hydration along crystal interfaces and cleavage planes with consequent differential volume change.

Above the weathered bedrock is a highly weathered layer, referred to as *structured saprolite*. This layer almost perfectly preserves the structures of the parent rock (Fig. 1-1*B*), but the rock has been reduced largely to clay and iron oxides, and its density is only about half that of the original rock. Most of the alkalis (Na, K) and alkaline earths (Ca, Mg) have been removed by leaching. More than 90 percent of this layer is iron, alumina, silica, and combined water. The colors are bright but varied. In places, especially where the parent rock is

Figure 1-1*B*.   In this photograph, the hammer marks the boundary between clayey, decomposed saprolite (structured saprolite) and firm but weathered bedrock below. Although they look alike in this photograph, the structured saprolite that perfectly preserves original rock structures is bright red clay. The bedrock is dull brown, tough rock.

basic and has a high content of iron-magnesium minerals, the saprolite is reddish brown. Where the parent material is granite, gneiss, or other acidic rock containing aluminous and siliceous minerals, the colors are reddish brown, light brown, yellow, purple, and white. The colors may occur as discontinuous lenticular bands or as irregular mottlings. The structured saprolite is commonly tens of feet thick, and may exceed 100 feet (33 m), especially in the South.

Above structured saprolite is a massive layer, generally less than 5 feet (1.5 m) thick. This material has about the same general appearance and feel as the structured saprolite except that it is massive and preserves little or no trace of the original structures. The boundary with the underlying structured saprolite is gradational through a zone no more than 1 or 2 cm thick. Quartz veins in structured saprolite end upward at this contact, but pieces of vein quartz are scattered in the massive layer above the veins. In places, a pebble layer at the base indicates local erosion. On hillsides, fragments of the vein quartz are distributed downhill from the upper reach of veins preserved in the structured saprolite. This layer has been mixed by frost heaving and creep, probably during the last Pleistocene glaciation.

The massive layer is at least 95 percent iron, aluminum, silica, and combined water. Some parts are snow white, but most are stained with iron oxides and are red, purple, brown, or yellow. In places the iron oxides are concentrated in nodules, layers, or cross-cutting veins.

Modern red and yellow Podzolic soils are developed on the massive saprolite (Fig. 1-1*A*). These relationships illustrate the confusion about the meaning of the term *soil*, for in this situation an ancient soil is parent material for a younger soil.

The massive saprolite on flat uplands is probably a relict from the Pleistocene when frost action was more vigorous than now, a hypothesis that would account for the sharp, smooth boundary at the base of the massive layer and the fact that the massive layer seems to thin southward from an average of about 5 feet (1.5 m) in Pennsylvania and Maryland to an average of only about 2½ (.75 m) feet in the Carolinas and Georgia. Local exceptions to this generalization may be attributed to local erosion and/or sedimentation.

Saprolite illustrates the common, and generally erroneous, concept that soil (in the agronomic sense) grades downward to bedrock. The concept is erroneous because, except for saprolite and other residuum, soils are generally the weathered upper layers of surficial deposits derived by weathering at one location and transported and deposited at a new location—like stream or glacial deposits. Saprolite is exceptional, as already mentioned, in being parent material for the modern (Late Pleistocene or Holocene) soil. Figures 1-1*B* and 1-2 illustrate some of the kinds of boundaries between weathered rock and saprolite.

VARIETIES OF SAPROLITE SHOWN ON THE MAP

Micaceous Residuum Without Much Quartz Sand (rsh)

On the Piedmont Plateau in the eastern United States, at least three kinds of Precambrian or Early Paleozoic metamorphic formations yield clayey saprolite that contains abundant relict mica but little quartz sand. The deposits are extensive on such parent rocks as schists, slate, and metamorphosed volcanic

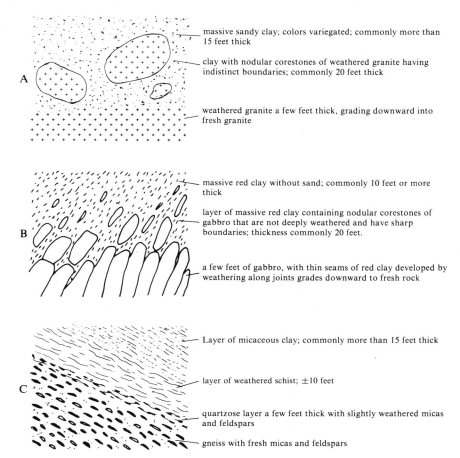

massive sandy clay; colors variegated; commonly more than 15 feet thick

clay with nodular corestones of weathered granite having indistinct boundaries; commonly 20 feet thick

weathered granite a few feet thick, grading downward into fresh granite

massive red clay without sand; commonly 10 feet or more thick

layer of massive red clay containing nodular corestones of gabbro that are not deeply weathered and have sharp boundaries; thickness commonly 20 feet.

a few feet of gabbro, with thin seams of red clay developed by weathering along joints grades downward to fresh rock

Layer of micaceous clay; commonly more than 15 feet thick

layer of weathered schist; ±10 feet

quartzose layer a few feet thick with slightly weathered micas and feldspars

gneiss with fresh micas and feldspars

Figure 1-2. Examples of kinds of boundaries between weathered rock and saprolite in granite (*A*), gabbro (*B*), and gneiss (*C*). The boundary between the weathered and fresh rock is gradational. In homogeneous and coarse-grained rocklike granite, the gradation around corestones consists of progressive change in degree of decomposition of the rock. In dense rocklike gabbro, the gradation around corestones is abrupt and the rounded remnants of unweathered rock decrease in size upward. Corestones of gneiss are uncommon.

rocks. The clay mineral is mostly kaolinite. Some samples contain small quantities of montmorillonite and, rarely, very small quanities of gibbsite.

Along the western base of the Sierra Nevada, similar saprolite has developed over metamorphosed sedimentary rocks, and in the Cascades clayey saprolite has developed over volcanic rock. Along the Pacific Coast north of San Francisco Bay, saprolite overlies weathered rocks of mixed types.

Clayey Residuum with Considerable Quartz Sand (rgr)

Granitic and gneissic rocks in the Piedmont Province are overlain by clayey saprolite containing considerable quartz sand. The quartz is relict from the parent rock. The deposits, as shown by the map, are only a little less extensive than those over schist. Most of the rocks here referred to as granitic contain considerable plagioclase as well as potassium feldspar. The granitic rocks are of igneous origin; the gneissic rocks are similar to the granites in mineral and chemical composition, but many or most of them are metamorphosed sedimentary rocks.

Saprolite developed over granite is typically massive and does not show much internal structure except for quartz veins. The veins are generally fragmented and are displaced on hillsides where mass wasting has caused downhill creep. The gneissic rocks, on the other hand, have well-developed structured saprolite. Both the granites and the gneisses have considerable biotite and muscovite, and flakes of altered mica occur in the clayey matrix, but the quantity is small compared to that in saprolite over schist.

Figure 1-3 contrasts weathering of granite in a semiarid and in a humid region. Analysis of saprolite over such rocks shows an increase in clay upward.

Clayey Residuum with Little Mica or Quartz (rga)

Saprolite on gabbro, basalt, and similar basic rocks consists chiefly of bright red, massive, kaolinitic clay on the Piedmont Plateau and in the Pacific Northwest. On the Piedmont Plateau these deposits are shown only in the Baltimore area, where they are moderately extensive. Small areas of similar deposits occur farther south on the Piedmont, particularly along the numerous Triassic dikes. The thickness of this saprolite is generally less than that over schistose or granitic rocks.

Saprolite over gabbro and related rocks, unlike that over schists or granite, grades sharply into fresh, dense rock, commonly within a quarter-inch (Fig. 1-2). The near-surface layers of this saprolite are massive and continuous. Beginning 5 to 10 feet (1.5 to 3 m) below the surface, rounded corestones of unaltered gabbro or diabase are embedded in the red clay, like the corestones in granite. At deeper levels over gabbro the clay extends downward in narrow bands along fissures that are connected by narrow cross-fractures containing clay. The intervening rock is fresh, but the masses are rounded by weathering as in the weathering of granite.

Red Clay (Terra Rossa) on Limestone (rls)

Red clay (rls on the map) is associated with numerous outcrops of limestones, caverns, sinks (karst topography), and large springs. The clay is massive and generally less than 10 feet (3 m) thick. Figure 1-4 illustrates an occurrence in Indiana.

In these red clays TiO^2 is increased as much as fortyfold over that of the limestone. This fact, together with the absence of structured saprolite in these clays, suggests that at least part of the clay is weathered, wind-blown dust.

Cherty Red Clay (rlc)

The deposits shown on the map as cherty red clay are similar to terra rossa but contain abundant chert derived from the parent limestone formation. They are found mostly in Missouri.

Figure 1-3. Weathered granite in a semiarid region. *Top:* (near Prescott, Arizona), and in a humid region, *Bottom:* (near Blowing Rock, North Carolina in the southern Blue Ridge). The proportion of decomposed granite around corestones in the humid regions is very much greater than in the semiarid regions, and the corestones become isolated as boulders as the decomposed material is eroded.

Figure 1-4. Residual clay on limestone in southern Indiana (near Marengo). The clay, marked by grass and weeds in this black-and-white photograph, is bright red (terra rossa) and contrasts sharply with the underlying bare gray limestone. This residuum is found south of the Wisconsinan glacial border. It probably includes admixed loess, but it is very much thinner than the residuum on the Piedmont Province.

Residuum on Triassic Sedimentary Formations in the Eastern United States (rtr)

The Piedmont Plateau has, in addition to the Triassic diabase already referred to, extensive beds of Triassic sandstone, shale, and conglomerate in structural basins now isolated from each other. Most of the conglomerate is quartz and sandstone containing feldspars that are decomposed at outcrops. As a consequence of the high percentage of quartz, the surficial materials are not as clayey or as conspicuous as those on the surrounding, older, metamorphic rocks.

Sandy Residuum (rs)

Deep weathering of the sandy coastal plain formations, especially in New Jersey and along the inner edge of the coastal plain in the southeastern states, has developed sandy residuum that in places develops dunes. The sand hills of New Jersey (Fig. 1-5) and the Carolinas are examples. Belts of sandy material among other deposits shown as loam (rl) would also be shown as sandy residuum on more detailed maps. Both the sandy residuum and loam form belts nearly parallel to the coast.

Clayey Residuum with Swelling Properties (rc)

Red and yellow residuum characterized by swelling clay (probably montmorillonite) has been developed by Tertiary and Quaternary weathering of poorly consolidated, shaly formations in the Gulf Coastal Plain. Formations containing swelling clay form ground that swells and cracks on wetting and drying. In Texas these

Figure 1-5.   Residuum on sandy coastal plain formations is commonly reworked into dunes, as in this view of New Jersey's pine barrens.

include Cretaceous formations in a belt extending from San Antonio through Austin to Dallas and some Eocene formations between the Rio Grande, Nueces, and Colorado (of the east) rivers. Swelling clay is also found in some of the Pleistocene deposits along the Gulf coast that are shown on the map as groundwater Podzol (gp). In Mississippi the ground with swelling clays is developed on Cretaceous formations in the northeast part of the state and on Tertiary formations east of Jackson.

These and other clayey deposits on the coastal plain formations are closely related to saprolite on the Piedmont Province. But because the coastal plain formations themselves were in part derived by erosion of saprolite that covered the Piedmont Province and other parts of the Appalachians, the weathering is less obvious than on the crystalline rocks of the Piedmont Province. The depth to the parent formations, which are not very well indurated, is generally less than 10 feet (3 m), and below that depth the formations are likely to retain limy fossil shells.

Loamy Residuum on Poorly Consolidated Coastal Plain Formations (rl)

Residuum shown as loam on the map, although mostly loam, ranges from sand to clay. Each textural type forms a belt roughly paralleling the coast and reflecting the lithology of the underlying weathered parent formation. The clay is mostly kaolinite; the comparative absence of swelling clay in the loam may be attributed to the paucity of volcanic-source rocks in the Cretaceous and Tertiary formations (except in the west Texas coastal plain). The depth of the deposits is generally less than 10 feet (3.1 m), and below that limy shells are preserved in the parent formations. Outcrops of the coastal plain formations are mostly confined to bluffs along undercut sides of streams or at wave-cut shores, like the Calvert Cliffs in Chesapeake Bay and the Atlantic Highlands in northeastern New Jersey.

Highly Weathered Late Tertiary and Quaternary Gravels (rg, Shown Incompletely)

Although these deposits are shown on the 1/7,500,000 map only in the north part of the Atlantic Coastal Plain, they are also present in the south part of the Atlantic Coastal Plain, the Gulf Coastal Plain, and on high terraces along the Mississippi and other rivers of the Middle West. The deposits cap uplands between rivers. Where the gravel is weathered, cobbles of schist and granite are altered to clay (Fig. 1-6), like the saprolite on the Piedmont Plateau. The gravels, generally less than 30 feet (9 m) thick, overlie older saprolite, showing that there have been several episodes of saprolite development.

Figure 1-6. Weathered Pleistocene gravel (Kansan?) about 200 feet (61 m) above the Susquehanna River at Havre de Grace, Maryland. All the cobbles and pebbles, except those of massive quartz, are so altered to clay that one has to look closely to identify the deposits as originally gravel.

Phosphatic Clay (rph)

Phosphatic clays occur mostly in Florida, although some, not shown on the map, are found near the coast of South Carolina. They consist of poorly sorted clay and phosphate pebbles or nodules in a sandy matrix locally cemented by iron oxide. Colors are white to brown and dark gray. The deposits are attributed to weathering of a shallow-water, marine deposit of Pliocene age originally containing apatite, $Ca^5(PO^4)^3(OH)$. Weathering alters the apatite to wavellite, $Al^2(PO^4)^2(OH)^3.5H^2O$. Paralleling this change, montmorillonite alters to kaolinite with loss of silica, which is probably the source of the abundant silica nodules in the underlying Miocene limestone. The phosphatic clays are important sources of phosphate for fertilizer, and the deposits are being extensively strip-mined. Figure 4-7 illustrates the stratigraphic relation of these deposits to old beach deposits and other features along the Florida shore.

Sandy or Silty Residuum (rsi)

Ancient and deep weathering of Paleozoic sandstone and shale in the south-central states has developed sandy or silty residuum; the deposits probably include loess. Thickness generally is less than 10 feet (3 m). The subsoil is neutral in reaction, with pH about 7.0. Figure 1-7 illustrates a typical occurrence in Kentucky.

Figure 1-7. Residuum formed by pre-Wisconsinan weathering of Pennsylvanian sandstone and shale in the western Kentucky coal basin. The residuum is brilliant red, in contrast to the somber colors of the underlying Pennsylvanian rocks.

THIN RESIDUUM: WEATHERED MATERIALS ON YOUNG GROUND

In contrast to the thick residuum of humid regions, residuum on the younger ground in semiarid and arid regions, as the map shows, is thin and only slightly weathered. The residuum commonly contains admixed sediment, especially eolian and/or alluvial silt and sand. The residuum is generally less than 3 feet (.91 m) thick and generally has a white subsoil of calcium carbonate (caliche).

Silt on Limestone (ls)

Silty residuum lies on and occupies the major part of the Edwards Plateau in central and western Texas. The landscape is rolling to hilly and from 2,000 to 3,000 feet (609 to 914 m) in altitude. The silt rests on hard, Cretaceous limestone. The plateau is deeply eroded, and limestone outcrops are extensive. Probably much of the silt is of wind-blown origin. Towards the east the silt is dark brown; westward, where precipitation is somewhat less, the silt is sandier and light brown or gray. The contact between the silt and underlying bedrock is generally sharp, suggesting that the deposits are in part transported. Rainfall ranges from 15 inches (38 cm) in the west to about 25 inches (63 cm) in the east, but evaporation rates are high and the effective moisture is much less than the precipitation records suggest. Moreover, the region is subject to frequent, severe droughts.

Sandy Residuum (ss)

Sandy Tertiary and older formations on the Great Plains, Wyoming Basin, and Colorado Plateau have a discontinuous, thin mantle of loose sand (Fig. 1-8), generally less than a foot thick, and containing fragments of the sandstone. Most of the sand is collected in open fissures in the underlying sandstone, and the scrubby vegetation on such surfaces is commonly aligned along the fissures as if artifically planted in rows. The sandstone units in places form flat-topped mesas; more commonly the surfaces are sloping and end up-dip in cuesta escarpments. Downslope the sandstones extend beneath overlying shale formations. Where the sand is thick, it generally develops dunes, most of which are too small to show on the map.

Shaly Residuum (sh)

Shaly residuum is developed on shale formations (mostly Cretaceous and Tertiary) but includes sandy deposits, like those labelled ss, and small areas of gravel. Typically the landscape characterized by shaly residuum consists of badland scarps and ridges with smooth erosion surfaces (pediments) sloping from their base (Fig. 1-9). Where precipitation is less than about 15 inches (38 cm) annually, the shale badlands weather into shale flakes in a fluffy surface layer about a foot thick. This active layer collects on steep badland hillsides; its fluffiness (density < 1.0) is due to swelling clay, probably montmorillonite, in

Figure 1-8. This sandy residuum is on an eroded surface of sandstone (Dakota Sandstone of Cretaceous age) on the semiarid Colorado Plateau in western Colorado. View northwest to the La Sal Mountains.

Figure 1-9. Cretaceous shale badlands (Henry Mountains region, Utah). The sandstone-capped mesas (left) are mantled by thin sandy residuum, like that shown in Figure 1-8. The badlands are mantled by a fluffy layer composed of shale flakes. At the base of the badland slopes are flat areas (pediments) eroded on the shale and mantled by a thin, compact layer of alluvium that grades to thick alluvium along the main streams.

the parent shale formation. The fluffy layer erodes several ways, depending on the kind of rainstorm. Some rains simply soak the layer, causing small slides and exposing bare shale. Harder rains may cause gullying, but during the dry period following such storms the gullies usually heal, partly by filling with shale flakes, partly by creep. The shaly residuum is, therefore, a comparatively active deposit.

As the shale badland hills are reduced in height, their slopes are reduced. Where relief on the badlands exceeds a certain minimum (in a particular precipitation zone), the angles of the hillside slopes remain about constant until erosion reduces the height of ridges to less than the width, after which the lowered hills become low, broadly convex mounds.

Sandy Gypsiferous Residuum (gyp)

Surficial deposits of gypsum mixed with sand or limy loam are shown on the map only in southeastern New Mexico and western Texas, where Permian salt and gypsum-bearing formations are extensive. Other local deposits, too small to show on the map, are found in Kansas (e.g., Big Basin), at the salt anticlines in west-central Colorado and east-central Utah (Fig. 1-10), and around many playas in the Basin and Range Province, including the Great Salt Lake desert, Utah, and Lake Lucero, New Mexico (at White Sands National Monument).

Clayey Residuum on Weathered Permian and Triassic Red Beds (c)

Deposits of red clayey residuum on the Great Plains are deceptively similar to saprolite, but their color is largely inherited from the parent Permian and Triassic rocks. The deposits include considerable loess. Subsoils are limy. The

Figure 1-10. In the western United States, gypsiferous formations are extensive and the ground is sandy and gypsiferous (light foreground). This view on the Colorado Plateau, shows gypsiferous ground on a salt anticline at Paradox Valley in western Colorado. The valley is a paradox because the main stream, the Dolores River (marked by a line of cottonwood trees in the middle distance), flows at right angles across Paradox Valley.

deposits are thin, generally less than 4 feet (1.2 m) and generally only a foot (.30 m) thick. Outcrops of bedrock are numerous. These deposits may be bordered by and confused with pediments mantled by alluvial silt also derived from the bedrock formations.

Pediments, especially on Cretaceous shale formations, develop at the base of badland scarps and are mantled by alluvium that is commonly about 6 inches (15 cm) thick. This is a dense, compact layer that dries like tough adobe. The surface is cut by numerous rills an inch or so deep, and a 30-year photographic record of such a surface in the Henry Mountains region, Utah, showed almost no change in the tiny rills of that surface. In other areas where annual precipitation averages more than about 15 inches (38 cm), shale badlands are likely to be only locally developed for they are quickly reduced by erosion.

Included with these areas are remnants of old erosion surfaces that are capped by alluvium or gravel and that form small mesas in the midst of the badlands.

STRATIGRAPHY OF RESIDUUM

Evidence is abundant, from coast to coast along the border of the Wisconsinan glacial drift and from that border southward, indicating that even the youngest saprolite is pre-Wisconsinan. Some saprolite is as old as Jurassic or earliest Cretaceous (e.g., the bauxite deposits of the Gulf Coastal Plain). The quite different kinds of alteration in the ancient soils (Fig. 1-1) have been described by Leighton and MacClintock (1930) as follows:

Surface
Layer 1: Modern soil
Layer 2: Chemically decomposed parent material; considerable clay developed by alteration of hard rocks
Layer 3: Leached of carbonates and other soluble salts and oxidized; otherwise little altered
Layer 4: Oxidized, but not leached and not otherwise altered; may contain some clay washed from above
Layer 5: Unaltered parent material

The diagnostic feature of the ancient soils is the chemical decomposition of hard rocks and their alteration to clay in Layer 2. Boulders or cobbles of granite may be so altered that they can be sliced by a knife. In contrast, deposits of the Wisconsinan glaciation and younger deposits are little weathered. The kind of alteration attributable to Wisconsinan and later weathering is the kind seen in Layers 3 and 4 of the ancient soils—that is, leaching of soluble salts and oxidation, but little or no alteration to clay.

On moraines and other deposits of Late Wisconsinan age, most stones, even in the upper layers, are firm and have only a very thin weathering rind. On Early Wisconsinan deposits most stones are firm, but many have an oxidized weathering rind an eighth- or quarter-inch thick. Of the stones on the surface or in the upper layers of the ancient soils, however, a high percentage are clayey, crumbly, or otherwise unsound—a fact of economic as well as of stratigraphic interest, for it limits the usefulness of the gravel.

Along with this contrast in weathering of Wisconsinan and pre-Wisconsinan deposits is a difference in the degree of erosion of the deposits. Whether in the eastern, central, or western parts of the United States, the moraines and other glacial deposits of Wisconsinan age retain the constructional geomorphic form they had when they were originally laid down by the ice or its meltwaters. The landforms have not been greatly modified by erosion. The pre-Wisconsinan deposits, however, are not only deeply weathered but are so eroded that the original geomorphic forms have been considerably modified or destroyed.

In the central United States around the Mississippi Embayment bauxite deposits have formed as a result of intensive weathering similar to that which produced the saprolite on the Piedmont Province. Rock formations older than the Wilcox Group of Eocene age were deeply weathered, and bauxite and related clayey sediments derived from the bauxitic saprolite are contained in the Eocene formations. This weathering occurred after the sediments of the Paleocene Midway Group were deposited and before the Wilcox Group was deposited.

Pre-Wisconsinan soils are known in many places in the semiarid and arid states too. These residual deposits commonly have an upper layer of reddish clay or silty clay that is generally no more than a foot thick, but a maximum of about 10 feet is known (in the Wasatch Mountains, Utah). The clay layers probably include wind-blown silt. The clay is leached of its carbonate and other soluble minerals.

Beneath the clay is a conspicuous white layer of calcium carbonate (caliche, see Chapter 2), commonly 2 to 3 feet (½ to 1 m) thick. It probably formed by weathering of wind-blown, carbonate-rich dust, added in small increments as desert-derived loess, especially in the desert southwest. The reality of such carbonate-rich dust is attested by the spots that collect on a wet windshield or window. To remove them requires washing with vinegar or other acid!

These western soils, although overlapped by Wisconsinan-age deposits, are mostly of Pleistocene age. The West is a young landscape and generally lacks soils that are Tertiary or older. However, some Late Cretaceous and Tertiary formations are attributed to erosion of saprolitic soils.

North of the Wisconsinan drift border, clayey alteration attributable to weathering is found only very locally in excavations extending below the drift. The unconformity at the drift border shows, too, in a pedologic unconformity. On the glacial drift the parent materials are fresh, hard rocks in the sand and gravel of the drift, and the modern soils on these deposits lack the layer of clay alteration. Similar soils are found south of the drift border on glacial outwash and other young deposits. But south of the drift, surfaces older than the Wisconsinan glaciation are extensive, and on these old weathered surfaces saprolite with red and yellow Podzols has developed. Depending entirely on accidents of erosion, the modern (Holocene) soil may be developed on unweathered parent material, on a layer that is oxidized but not leached, on a layer that is both oxidized and leached, or on a layer where alteration has progressed to clay.

At many places in the Tennessee Valley, Shenandoah Valley, and valleys in the Piedmont Province, saprolite is unconformably overlain by alluvium that probably is late Pleistocene in age.

Near the Fall Line, along the Kennedy Expressway north of Baltimore, the Patuxent Formation (Early Cretaceous) overlaps saprolite and is in part derived by reworking of the saprolite (Fig. 1-11). Locally this particular saprolite and the

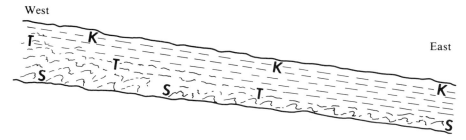

Figure 1-11. During construction of the Kennedy Expressway near Baltimore, an excavation about 1,000 feet (305 m) long exposed saprolite (S) reworked into transitional colluvial material (T) and overlain by the Cretaceous Patuxent Formation (K). The lower part of the Patuxent grades into the colluvial layers that, in turn, grade laterally to saprolite.

sedimentary bed derived from it grade so completely into each other that the boundary between them is arbitrary.

Some of the saprolite is therefore older than this Lower Cretaceous formation. The lithology of the Tertiary formations on the coastal plain suggest that they, too, were derived in large part from saprolite. The accumulated evidence shows that saprolite was developed at various episodes during the Tertiary as well as earlier in Cretaceous or Jurassic times and later in Early and Middle Pleistocene.

Although saprolite, like modern soils, has layers parallel to the ground surface, part of the alteration is attributable to groundwater moving laterally along the weathered rock beneath the structured saprolite. Commonly this contact is wet, and where it is truncated by a hillside or road cut, it is the site for springs.

Other evidence that saprolite may be deepened by groundwater action is provided by the geometry of some deposits. Along Gunpowder Falls near Baltimore, for example, saprolite is 50 feet (15 m) deep under the stream bed, but a short distance away the stream turns into a bedrock channel that is 30 to 40 feet (9 to 12 m) higher than the base of the saprolite. Similar relationships are known in the Tennessee and Shenandoah valleys.

Figure 1-12 illustrates some principles of soil stratigraphy. The Wisconsinan soil on the hill is the parent material for the Holocene soil, and the pre-Wisconsinan soil is the parent material for the Wisconsinan soil. It must be stressed further that not only are the three soils separable stratigraphically, but each of the older soils was fully formed *before* the overlapping sand was deposited. Weathering during Wisconsinan and Holocene times has contributed little to the development of the pre-Wisconsinan profile. The differences in the degree and kind of weathering represented by these soils are not due to the factor of *time* per se, but are attributable to differences in the processes that operated during the development of each. For example, in many places the depth of the old soils is very much greater than the depth to which moisture now penetrates.

It is exceptional to find Holocene, Wisconsinan, and pre-Wisconsinan soils in one section. More commonly the old soils have become the parent materials for the younger ones and the soil profiles are superimposed (Fig. 1-12). Moreover,

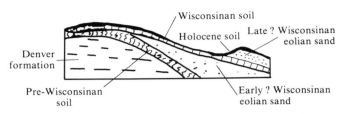

Figure 1-12. Superimposed weathering profiles. Depending on accidents of erosion, the layers of an ancient (pre-Wisconsinan) soil may become the parent material for a Late Pleistocene or Holocene soil. The section illustrates how superimposed weathering profiles can be traced laterally to unweathered parent materials that separate soils of different ages. At right, three weathering profiles of different ages are separated by unweathered deposits datable by fossils and artifacts. Towards the left the Late Wisconsinan sand thins and the Holocene weathering profile is superimposed on the Wisconsinan soil. Further left, the Early Wisconsinan sand thins and the combined weathering profiles are superimposed on one that is pre-Wisconsinan. (After Hunt, 1954)

the parent materials of a young soil may be either the upper or lower horizon of an older one.

Accordingly the pre-Wisconsinan, Wisconsinan, and Holocene weathering profiles can be completely superimposed, giving rise to many very different combinations of the weathering effects. Three examples help illustrate the variety of possible combinations:

1. Holocene weathering may have developed on the upper layers of a Wisconsinan soil that was superimposed on the upper or lower layers of a still older weathering profile; or,
2. Holocene weathering may have developed on the lower layers of a Wisconsinan soil that was superimposed on the upper (or lower) layers of a still older weathering profile; or,
3. Erosion may have removed a Wisconsinan weathering profile, and the Holocene weathering may have developed on a layer of the pre-Wisconsinan soil.

Such complications occur in every part of the country that was not buried under Wisconsinan glacial deposits. Figure 1-13 illustrates the stratigraphic relations of such residuum near the Wisconsinan drift border in northern Pennsylvania.

REFERENCES CITED AND ADDITIONAL BIBLIOGRAHY

Altschuler, Z. S., E. J. Dwornik, and H. Kramer, 1963, Transformation of montmorillonite to kaolinite during weathering, *Science* **141:**148-152.

Borns, H. W., Jr., and H. W. Allen, 1963 Pre-glacial residual soil in Thomaston, Maine, *Jour. Sed. Petrology* **33:**675-679.

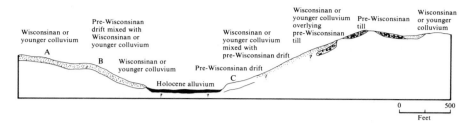

Figure 1-13. Idealized section of surficial deposits and soil in an area south of the Wisconsinan drift border in northern Pennsylvania showing dissected uplands mantled by pre-Wisconsinan or younger colluvium and mixtures of these deposits. The pre-Wisconsinan deposits are strongly weathered, are yellowish red, and contain much clay. The Wisconsinan deposits have well-developed soils but are not brightly colored and have little clay attributable to weathering. Holocene deposits are only slightly weathered. (After Denny and Lyford, 1963)

Cleaves, E. T., 1968, Piedmont and Coastal Plain geology along the Susquehanna aqueduct, Baltimore to Aberdeen, Maryland, *Maryland Geol., Survey Rept. Inv.* **8**, 45p.

Cline, M. G. and others, 1955, Soil survey of the Territory of Hawaii, *U.S. Dept. Agriculture Soil Survey Ser. 1939*, no. 25, 644p.

Crocker, R. L., 1946, Post-Miocene climatic and geologic history and its significance in relation to the genesis of the major soil types of South Australia, *Australia Council for Sci. and Ind. Res. Bull. 193*, 56p.

Denny, C. S., 1956, Surficial geology and geomorphology of Potter Co., Pennsylvania, *U.S. Geol. Survey Prof. Paper 288*, 72p.

Denny, C. S., and W. H. Lyford, 1963, Surficial geology and soils of the Elmira-Williamsport region, New York and Pennsylvania, *U.S. Geol. Survey Prof. Paper 379*, 60p.

Droste, J. B., 1956, Alteration of clay minerals by weathering in Wisconsin till, *Geol. Soc. America Bull.* **67**:911-918.

Droste, J. B., and J. C. Tharin, 1958, Alteration of clay minerals in Illinoian till by weathering, *Geol. Soc. America Bull.* **69**:61-68.

Dryden, L., and C. Dryden, 1946, Comparative rates of weathering of some heavy minerals, *Jour. Sed. Petrology* **16**:91-96.

Goldich, S. S., 1938, A Study in Rock Weathering, *Jour. Geol.* **46**:17-58.

Gordon, M., Jr., J. L. Tracey, Jr., and M. W. Ellis, 1958, Geology of the Arkansas bauxite region, *U.S. Geol. Survey Prof. Paper 299*, 268p.

Grim, R. E., 1953, *Clay Mineralogy*, McGraw-Hill, New York, 384p.

Harris, R. C., and J. A. S. Adams, 1966, Geochemical and mineralogical studies on the weathering of granitic rocks, *Am. Jour. Sci.* **264**:146.

Hawley, J. W., 1965, Geomorphic surfaces along the Rio Grande valley from El Paso, Texas to Caballo Reservoir, New Mexico, *New Mexico Geol. Soc. Guidebook 11*, pp. 188-197.

Hunt, C. B., 1954, Pleistocene and Recent deposits in the Denver area, Colorado, *U.S. Geol. Survey Bull. 966-C*, pp. 91-140.

Hunt, C. B., 1972, Geology of Soils, W. H. Freeman and Co., New York, 344p.

Hunt, C. B., 1972, Thirty year photographic record of a shale pediment, Henry Mountains, Utah, *Geol. Soc. America Bull.* **84**:689-696.

Hunt, C. B., and V. P. Sokoloff, 1950, Pre-Wisconsin soil in the Rocky Mountain region, *U.S. Geol. Survey Prof. Paper 221-G*, pp. 109-121.

Hunt, C. B., T. W. Robinson, W. A. Bowles, A. L. Washburn, 1966 Hydrologic Basin, Death Valley, California, *U.S. Geol. Survey Prof. Paper 494-B*, 138p.

Jackson, M. L., and G. D. Sherman, 1953, Chemical weathering of minerals in soils, *Adv. Agronomy,* **5:**219-318.

Kay, G. F., and J. N. Pearce, 1920, Origin of gumbotil, *Jour. Geology* **28**(2):89-125.

Kaye, C. A., 1951, Some paleosols of Puerto Rico, *Soil Sci.* **71:**329-336.

Kaye, C. A., 1967, Kaolinization of bedrock of the Boston, Massachusetts area, *U.S. Geol. Survey Prof. Paper 575-C,* pp. C165-172.

Keller, W. D., 1955, *The Principles of Chemical Weathering,* Lucas Bros., Columbia, Mo. 88p.

Keller, W. D., 1966, Geochemical weathering of rocks—source of raw materials for good living, *Jour. Geol. Education* **14**(1):17-22.

King, P. B., 1949, The floor of the Shenandoah Valley, *Am. Jour. Sci.* **247:**73-93.

Leighton, M. M., and P. MacClintock, 1930, Weathered zones of the drift sheets of Illinois, *Jour. Geol.* **38:**28-53.

Leverett, F., 1898, Weathered zones and soils (Yarmouth and Sangamon) between drift sheets, *Geol. Soc. America Bull.* **65**(5):369-383.

Marshall, C. E., 1964, *The Physical Chemistry and Mineralogy of Soils,* Wiley, New York.

Mason, B., 1958, *Principles of Geochemistry,* Wiley, New York, 310p.

Mertie, J. B., Jr., 1959, Quartz crystal deposits of southwestern Virginia and western North Carolina, *U.S. Geol. Survey Bull. 1072-D,* pp. 233-298.

Mohr, E. C. J., and F. A. Van Baren, 1954, *Tropical Soils,* Interscience, New York, 498p.

Nikiforoff, C. C., 1943, Introduction to paleopedology, *Am. Jour. Sci.* **241:**194-200.

Pettijohn, F. J., 1941, Persistence of heavy minerals and geologic age, *Jour. Geol.* **49:**610-625.

Rankama, K., and Th. G. Sahama, 1950, *Geochemistry,* University of Chicago Press, Chicago, 912p.

Richmond, G. M., 1962, Quaternary stratigraphy of the La Sal Mountains, Utah, *U.S. Geol. Survey Prof. Paper 324,* 135p.

Ross, C. S., 1943, Clays and soils in relation to geologic processes, *Washington Acad. Sci. Jour.* **33:**225-235.

Ruhe, R. V., R. B. Daniels, and J. G. Cady, 1967, Landscape evolution and soil formation in southwestern Iowa, *U.S. Dept. Agriculture Tech. Bull. 1349,* 242 p.

Simonson, R. W., 1954, Identification and interpretation of buried soils, *Am. Jour. Sci.* **252:**705-732.

Thorp, J., W. M. Johnson, and E. C. Reed, 1950, Some post-Pliocene buried soils of central United States, *Jour. Soil Sci.* **2**(pt. 1):1-19.

Valentine, K. W. G., and J. D. Dalrymple, 1976, Quaternary buried paleosols: a critical review, *Quaternary Res.* **6:**209-222.

Van Houten, F. B., 1961, Climatic significance of red beds, in A. E. M. Nairn, ed., *Descriptive Paleoclimatology,* Interscience, New York, pp. 89-139.

van Olphen, H., 1963, *An Introduction to Clay Colloid Chemistry,*, Interscience, New York, 301p.

Watson, T. L., 1902, Preliminary report on a part of the granites and gneisses of Georgia, *Georgia Geol. Survey Bull. 9 A,* pp. 292-356.

Yaalon, D. H., ed., 1971, *Paleopedology—Origin, Nature, and Dating of Paleosols,* Internat. Soc. Soil Sci. and Israel Universities Press, Jerusalem, 350p.

OTHER RESIDUAL DEPOSITS

ORGANIC DEPOSITS

Marsh, Swamp (m), and Peat (black triangle)

Poorly drained areas in humid climates collect water and support lush vegetation. The term *marsh* is generally applied to those wet areas having grasses, sedges, and other small plants; the term *swamp* usually applies to wet areas having trees (Figs. 2-1, 2-2, and 2-3). Either marshes or swamps may develop into peat bogs. Clayey or silty layers beneath such poorly drained ground are without oxygen, and that anaerobic condition weathers the mineral matter to sticky mud known as gley.

Peat is the partly carbonized organic residue produced by decomposition of roots, trunks of trees, twigs, seeds, shrubs, grasses (reeds), ferns, mosses, and other vegetation. The decomposition is slowed because the ground is saturated with water so that oxygen is excluded. Peat contains less than 4 percent inorganic matter, but it grades to muck, which is mud containing a high

Figure 2-1. Marsh ground (wetlands) on the Mississippi River delta.

Figure 2-2. A Florida swamp, characterized by cypress and Spanish moss.

Figure 2-3. A swamp in Maine, characterized by spruce.

percentage of organic matter. If the carbonized organic material burns freely when dry, it is usually classed as peat. Peat bogs are common in the glaciated parts of the country, including alpine uplands, and they also occur in the southeastern United States in limestone sinks, oxbow lakes along river floodplains, and places near the Atlantic and Gulf coasts.

The composition of peat varies greatly depending on the kind of vegetation and the quantity and kind of admixed sediment. A very rough approximation would be: carbon, 55 percent; nitrogen, 2 percent; hydrogen, 5 percent; oxygen, 33 percent.

Peat is lighter than water, and the specific gravity may be as low as 0.25. The texture depends on the kinds of plants. Four kinds of peat are: (1) turfy peat composed of slightly decomposed mosses; (2) fibrous peat composed of grasses, sedges with admixed leaves, or even wood; (3) earthy or woody peat, which dries into earthlike masses; and (4) pitchy peat, the heaviest kind, which breaks with smooth, lustrous fractures. Colors range from yellow through shades of brown to black. The heat value of moisture-free peat is about half that of bituminous coal. In the United States peat is used chiefly for gardening; its usefulness is its content of nitrogen and humus, which fertilizes soil, serves to loosen it, and holds moisture.

Related organic deposits include rice paddies, which are extensive in the Mississippi River alluvial plain and delta, mangrove swamps along the Florida coast, and the algal and other marshes at the zone of springs bordering around desert playas.

The paleontological record formed by remains of vegetation contained in surficial deposits may be found as an organic deposit that formed in marshes or swamps, or as pollen mixed with inorganic sediment. Figure 2-4 illustrates how fossil vegetation records vertical changes in the stratigraphy of deposits and how, by tracing the floras laterally, the vegetation records such changes in the environment.

Opinions differ about whether the latitudinal vegetation zones were crowded together into narrow bands as the Pleistocene glaciers advanced southward. Probably they telescoped, but to what degree is uncertain.

Sandy Coastal Ground with an Organic Layer over a Shallow Water Table (gp)

In sandy ground in humid climates, leaching may be intensive enough to extend all the way to the water table. Where the water table is shallow and fluctuates, the organic matter washing downward may accumulate and be preserved in a

Ground surface	*spruce and fir*	*alders and willows*	*sedges*	*ice retreat* ⟶ *mosses and algae*	*barren ground*
	debris of alders, willows	debris of sedges	debris of mosses, algae	barren	
	debris of sedges	debris of mosses, algae	barren		
	debris of mosses, algae	barren			
	barren				

Figure 2-4. Plant succession and the stratigraphy of the plant record in front of a retreating ice sheet. The top row shows a common succession of plant zones away from the front of the ice. The plants migrate towards the ice as the front retreats, and the buried plants contain a fossil record of the former position of each zone. The fossil material may be twigs, leaves or needles, cones, seeds, and, most important for the paleontological record, pollen.

zone kept moist by capillarity above the fluctuating water level. Such ground undergoes acid leaching and is referred to as groundwater Podzol (Fig. 2-5). As shown on the map, these deposits are extensive on flat, sandy stretches of the coastal plain in the southeastern United States. Similar acid-leached, poorly drained ground occurs in the glaciated northeastern and north-central United States, but these areas are too small to show on the map.

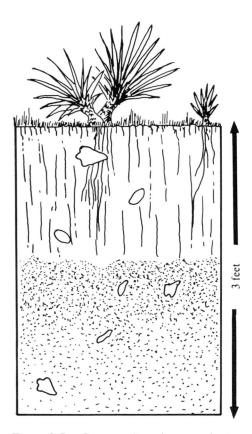

Figure 2-5. Cross section of a groundwater Podzol containing artifacts. At the top is an ash gray sandy layer 1-2 inches (2.5-5.1 cm) thick. Below this is a light gray or almost white leached zone, commonly about 18 inches thick. In the subsoil, 20-30 inches (50.8-76.2 cm) below the surface, is a dark layer impregnated with organic matter. The artifacts, which occur at all levels, are alike and represent a preceramic occupation. Locally the groundwater Podzols are overlain by shell mounds and trash piles of prehistoric Indians and contain pottery shards and arrow points. The groundwater Podzol is estimated to date from the Early Holocene. (After Hunt and Hunt, 1957)

Clinker (kk)

Clinker, formed by the burning of highly volatile lignite or subbituminous coal beds, could be considered a residual deposit. It is described in Chapter 9.

IMPORTANT RESIDUAL DEPOSITS NOT SHOWN ON THE MAP

Cave Deposits

Solution caverns are numerous in limestone terrain, in both the eastern and the western United States, and their stratigraphy is surprisingly alike at different latitudes and at different altitudes. Commonly, gravel is at the base, recording a stage of substantial water flow. Overlying this is clay or ochre deposited when the flow diminished. This in turn is capped by stalagmites, deposited when flow had diminished to a mere drip from the roof, and overlying the stalagmites is dust.

Remains of Pleistocene animals are common in the stalagmite and lower layers; equally generally, the layers above the stalagmite have yielded remains of a modern fauna. Early-man artifacts, when present, may underlie the stalagmite. The layers overlying the stalagmite contain artifacts of Holocene occupations. The bibliography lists some cave descriptions, including many that have important archeological remains.

Other kinds of caves include those in lavas where the lava surface cooled and crusted over a lava river, which could flow from under the crust. One such lava cave in New Mexico, south of Grants, has perennial ice. The floors of most lava caves are roughened by blocks of lava fallen from the roof. Alcoves are abundant in the cliff-forming sandstone on the Colorado Plateau; they are floored with loose sand and sandstone slabs fallen from the roof. Most alcoves formed where perched groundwater dissolved the cement of the sandstone, enabling wind to blow away the loosened grains. Many of these became utilized as shelters by prehistoric Indians, as at Mesa Verde National Park.

Caliche

Caliche is a Spanish-American term for certain kinds of crusty deposits formed by the accumulation of salts. Most of them contain calcium carbonate, but in some places they contain calcium sulfate or other salts. They are characteristic of the deserts and semiarid lands of the West, where they formed at various times during the Quaternary epoch. At least four quite different kinds can be distinguished.

First, the lime-enriched layers, caliche in "soils" in the western states, are the most extensive. In arid and semiarid regions, none of the small amount of rain that penetrates the ground reaches the water table; the water that does seep into the ground is simply held by capillarity until it evaporates. As a result, the soluble salts in the upper layer of the soil, especially the carbonates and sulfates that were deposited there as wind-blown silt, are dissolved and redeposited in a lime-enriched layer. These carbonate layers may be as much as 20 feet (6 m) thick under some pre-Wisconsinan soils.

One of the more complete studies of caliche has been made in New Mexico where Gile and others (1966) found orderly, age-correlative sequences in sandy and gravelly ground (Fig. 2-6). The sequence reported on gravelly ground is as follows.

Middle and Late Holocene caliche occurs as filaments and flakes on pebbles, forming thin, discontinuous coatings 6 to 18 inches (15 to 45 cm) below the surface. In places these can be found only by testing with acid. Early Holocene caliche is continuous between pebbles, but not hardened; coastings on pebbles are noticeable. Late Pleistocene caliche is continuous and partly cemented; its permeability is sharply reduced, and it is considered a clogged layer. Older, but still Pleistocene caliche, develops laminar structure above the clogged layer and has firmly cemented layers.

A second kind of caliche is the result of surf action at the shorelines of Pleistocene lakes (see Chapter 6). Such deposits cement gravels or other shoreline deposits forming beachrock.

A third kind of caliche deposit formed where water rose by capillarity above the water table. Like the shoreline deposits, most of these deposits are relics of times when the water table was higher than it is today.

A fourth kind of caliche is represented by tufa, or a thick layer of travertine, at springs (see Chapter 9), especially at warm springs like those in Yellowstone Park and in Death Valley. Huge mounds of travertine have been deposited by springs from the limestone formations along the Lower Granite Gorge of the Grand Canyon in Arizona.

Desert Pavement

A special kind of stony ground, known as desert pavement, is widely distributed on gravel fans and gravel terraces in arid and semiarid parts of the country. Desert pavement consists of a single layer of closely spaced stone (Fig. 2-7), which may be rounded or angular, and which overlies a vesicular layer of sand and silt. The sand and silt in turn overlie the fan or terrace gravels.

The stones collect at the surface by some kind of sorting action, apparently due to heaving by frost, precipitation and solution of salts, and wetting (with swelling) and drying of clays. The sand and silt underlying the pavement is probably in part of eolian orgin, because in most places comparable fine material does not occur in the parent gravels. A mixture of stones and silt does not produce a mix resembling the parent gravels.

In general, within a particular region, the thickness of the silt layer underlying the pavement increases with increasing age of the surface on which it occurs, and the angularity of the stones increases with the age of the pavement because of advanced weathering and disintegration (Fig. 2-8).

How desert pavements develop such smooth surfaces remains a problem. Denny (1965), who studied desert pavement on fans in the Amargosa Desert east of Death Valley, has suggested a mechanism involving lateral creep (pp. 19-21): "A lateral movement . . . is indicated by the miniature terraces, and this movement appears to be caused by the flowage of the silt on which the fragments rest. When saturated with water, the silt becomes plastic and tends to flow downslope."

Figure 2-6. Development of laminar caliche in the southwestern deserts. Shrubs are widely spaced; organic matter is concentrated around them, and the intervening ground is bare. (Chronology based on Gile et al., 1965; Gile, 1966; Hawley and Gile, 1966)

A. Mid-Holocene

A. Mid-Holocene caliche is barely visible; calcium carbonate coats sand grains and minute cracks.

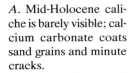

B. Late Wisconsinan caliche is clearly visible in a zone 5-6 inches (12.7-15.2 cm) thick and 4-6 inches (10.2-15.2 cm) below the surface.

B. Late Wisconsinan

C. Older Wisconsinan caliche is a conspicuous layer and slows downward percolation of water.

C. Older Wisconsinan

D. Early Wisconsinan and and older caliche has almost completely plugged downward percolation of water, and a laminar layer has formed across the top of the plugged layer.

D. Early Wisconsinan and older

E. Pre-Wisconsinan caliche may develop laminar layers 4-5 feet (1.2-1.5 m) thick over 2 feet (0.61 m) or more of nodular, fissure, and disseminated calcium carbonate. The laminated layers are commonly overlain by a foot or more of sand.

E. Pre-Wisconsinan

Figure 2-7. Desert pavement consists of a smooth layer of closely set stones. Such surfaces can be formed by wind removing fine particles, but the most extensive surfaces involve stones being heaved upward out of a surrounding matrix of eolian dust by frost, crystallization of salts, or swelling of clay. This is indicated by the common occurrence of 1-10 inches (2.5-25.4 cm) of fine silt without stones just beneath the pavement. There is little or no sand like that mixed in the underlying fan gravels. (Photograph by John Stacy, U.S. Geological Survey)

Rock (Desert) Varnish

As the name implies, desert varnish is best developed, or at least most conspicuous, in desert regions, but rock stain by iron and manganese oxides is by no means restricted to such areas; rock surfaces in humid regions are also stained. The stained surfaces may be the top or sides of isolated individual stones, or they may be vertical or overhanging cliffs or other surfaces moistened by seeps or splashed by rivers. Water is needed to transport iron and manganese to the surface, and in humid regions the deposits may form on railroad cuts or in wet tunnels in a few years. In arid regions, too, desert varnish is forming at live seeps. The problem about rock stain in deserts is how it became deposited so extensively on surfaces that are rarely wet, like the cliffs of the canyon lands on the Colorado Plateau and the expansive desert pavements on gravel fans in the Basin and Range Province.

Although varnish is being deposited today in humid regions and at seeps in semiarid and arid regions, the widespread deposits on surfaces in arid regions are clearly relics of past environments. Abundant stratigraphic evidence through-

Figure 2-8. "Ashtray weathering" of diorite porphyry on the Colorado Plateau (Henry Mountains area, Utah). This sort of weathering, as if by solution yet in semiarid and supposedly alkaline ground, may be due to acid clay. Figure 3-13 shows similar pitting on a very large boulder of the porphyry. No pattern of microclimate has been discovered to explain such pitting of seemingly insoluble rock.

out the Southwest and in other arid or semiarid parts of the world indicates that very little of the stain has been deposited in the last 2,000 years. Stone artifacts and stone structures of that period in the Southwest are not stained except where subjected to excess moisture. Early Holocene and older stone artifacts and structures are commonly stained. These generalizations are based on relationships at many (thousands, perhaps) archaeological sites in the Southwest. The same appears true in other arid regions. Evidence from Egypt indicates practically no deposition of desert varnish there in 2,000 years, slight deposition in 5,000 years, and dark stain on older stonework. The most comprehensive and definitive work on rock varnish is by Dorn and Oberlander (1982).

Weathering of Sulfide Desposits

The weathering of metal sulphide deposits develops orderly layers, analogous to those in deep residuum, but most such deposits are in the arid and semiarid West where residuum generally is thin. The weathering of sulfide minerals produces sulfuric acid, and as a consequence of its potency, the rocks are altered to great depth (Fig. 2-9). The differences in value of the ore at different depths are major, too, and have a substantial impact on the methods and economics of the mining operation. Most of the big copper deposits of the Southwest are layered about as illustrated in the figure.

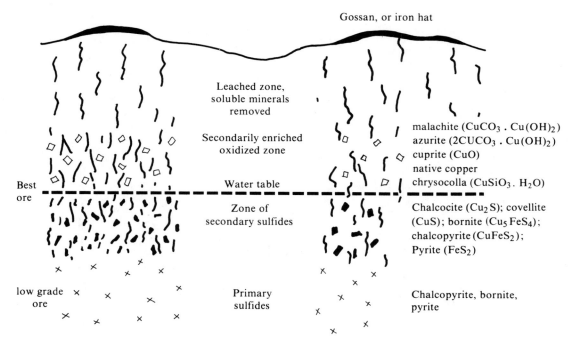

Figure 2-9. Weathering profile of a deposit of sulfide minerals, as in a disseminated copper deposit.

Turquoise

Turquoise, a highly prized gemstone occurring in the weathered, surficial layers of metal deposits containing phosphatic minerals, has been widely sought since prehistoric times, both in the southwestern United States and in the Middle East and eastward in Asia to Mongolia.

Turquoise is a hydrous phosphate of aluminum containing copper, which imparts the color. In the United States most of the producing mines are in the Basin and Range Province and the southern Rocky Mountains in Colorado. The matrix is generally deeply weathered, probably by sulphuric acid solutions analogous to those that altered the upper parts of lode deposits. The phosphorous has probably been derived by weathering of apatite in the parent rock, generally a volcanic or intrusive rock, and is commonly associated with various clay minerals, iron oxides and hydroxides, and other products of weathering.

Patterned Ground

Patterned ground (Washburn, 1973) refers to more or less symmetrical ground forms of stones, arranged as circles, polygons, nets, stripes, and steps, that are characteristic of, but not restricted to, ground subject to severe frost heaving. Patterned ground is common on mountain tops in the western United States (Fig. 2-10), and there are scattered localities in the eastern United States near the border of the glacial drift. The mechanisms that produce the ground patterns are not fully understood, but their great extent in cold regions indicates

Figure 2-10. A form of patterned ground, sorted polygons. These developed on Precambrian rocks on a treeless summit of the Uinta Mountains, Utah. Altitude approximately 12,000 feet (3,658 m).

frost action is clearly a major factor. However, very similar ground patterns in Death Valley (Hunt and Washburn, 1966) and other southwestern desert areas, where freezing temperatures are rare, must be attributed to salt heaving.

REFERENCES CITED AND ADDITIONAL BIBLIOGRAPHY

Braun, E. L., 1955, The phytogeography of unglaciated eastern United States and its interpretation, *Bot. Rev.* **21:**297-375.

Bretz, J. H., 1956, Caves of Missouri, Missouri Div. Geol. Survey and Water Resources [Rept.] 39, 2nd ser., 490p.

Bretz, J. H., and S. E. Harris, Jr., 1961, Caves of Illinois, Illinois Geol. Survey Rept. Inv. 215, 86p.

Bretz, J. H., and C. L. Horberg, 1949, Caliche in southeastern New Mexico, *Jour. Geology* **57:**491-511.

Bryan, K., 1941, Correlation of the deposits at Sandia Cave, New Mexico, with glacial chronology, in Evidences of early occupation in Sandia Cave, New Mexico, and other sites in the Sandia-Manzano region, *Smithsonian Misc. Colln.*, F. G. Hibben, **99**(23):45-64.

Bryan, K., 1950, Geological interpretation of the (Ventana Cave) deposits, in *The stratigraphy and archeology of Ventana Cave*, Arizona, E. W. Haury, Ariz. Univ., Tucson and New Mexico Press, Albuquerque, pp. 75-125.

Cannon, H. L., 1960, The Development of Botanical Methods of Prospecting for Uranium on the Colorado Plateau, *U.S. Geol. Survey Bull. 1085-A*, pp.1-50.

Carlisle, D., and G. B. Cleveland, 1958, Plants as guides to mineralization, *Calif. Div. Mines Spec. Rept. 50*, 31p.

Davies, W. E., 1950, The caves of Maryland, *Maryland Geol. Survey Bull. 7*, 75p.

Davies, M. B., 1963, On the theory of pollen analysis, *Am. Jour. Sci.* **261:**897-912.

Deevey, E. S., Jr., 1949, Biogeography of the Pleistocene, *Geol. Soc. America Bull.* **60:**1315-1416.

Deevey, E. S., Jr., 1958, Bogs, *Sci. Am.* **199**(4):114-122.

Denny, C. S., 1965, Alluvial fans in the Death Valley Region California and Nevada, *U.S. Geol. Survey Prof. Paper 466,* pp. 16-21.

Dorn, R. I., and T. M. Oberlander, 1982, Rock varnish, *Prog. Physical Geog.* **6**(3):317-367.

Davies, W. E., 1970, Karstlands and Caverns (maps), in *National Map Atlas*, U.S. Geological Survey, Washington, D.C., p. 77.

Gile, L. H., F. F. Peterson, and R. B. Grossman, 1966, Morphological and genetic sequences of carbonate accumulation in desert soils, *Soil Sci.* **101:**347-360.

Hanson, H. P., 1947, Postglacial forest succession, climate, and chronology in the Pacific Northwest, *Am. Philos. Soc. Trans.* **37**(pt.1), 130p.

Harrington, M. R., 1933, Gyspum Cave, Nevada, *Southwest Museum Paper 8.*

Hawley, J. W., and L. H. Gile, 1966, Landscape evolution and soil genesis in the Rio Grande region, southern New Mexico, *Friends of Pleistocene 11th Annual Field Conference Guidebook,* 74p.

Hayes, P. T., and B. T. Gale, 1957, Geology of the Carlsbad Caverns east quadrangle, New Mexico, *U.S. Geol. Survey Geologic Quad. Map CQ 98.*

Hibben, F. C., 1941, Evidences of early occupation in Sandia Cave, N. Mex., and other sites in the Sandia-Manzano Mountain region, *Smithsonian Misc. Colln.* **99**(23):1-64.

Hibben, F. C., 1955, Specimens from Sandia Cave (N. Mex.) and their possible significance, *Science* **122:**688-689.

Hunt, C. B., and A. P. Hunt, 1957, Stratigraphy and archaeology of some Florida soils, *Geol. Soc. America Bull.* **68:**797-806.

Hunt, C. B., and Washburn, A. L., 1966, Patterned ground in Hunt and others, Hydrologic basin, Death Valley, California, *U.S. Geol. Survey. Prof. Paper 494-B,* p. B104-B130.

Jennings, J. D., 1957, Danger Cave, *Am. Archaeol. Soc. Mem.* **32** (n. 2, pt. 2), 328p.

Jennings, J. N., 1971, *Karst* MIT Press, Cambridge, Mass., 252p.

Martin, P. S., 1958, Pleistocene ecology and biogeography of North America, Chap. 15 in pt. 2 of Hubbs, C. L., ed., *Zoogeography*, Am. Assoc. Adv. Sci. Pub. No. 51, p. 375-420.

Meinzer, O. E., 1927, Plants as indicators of ground water, *U.S. Geol. Survey Water-Supply Paper 577,* 95p.

Pogue, J. E., 1915, The turquoise: a study of its history, mineralogy, geology, ethnology, archeology, mythology, folklore, and technology, *Natl. Acad. Sci. Mem.* **12** (pt. 2, mem. 3), 162p.

Poltzger, J. E., and J. E. Otto, 1943, Post-glacial forest succession in northern New Jersey as shown by pollen records from 5 bogs, *Am. Jour. Botany* **30:**83-87.

Robinson, T. W., 1958, Phreatophytes, *U.S. Geol. Survey Water-Supply Paper 1423,* 89p.

Stock, C., 1931, Problems of antiquity presented in Gypsum Cave, Nevada, *Sci. Monthly* **32:**22-32.

Stone, R. W., 1932, Pennsylvania Caves, *Pennsylvania Geol. Survey Bull. G 3,* 141p.

Thorp, J., 1949, Effects of certain animals that live in soil, *Sci. Monthly* **68:**180-191.

Waksman, S. A., 1952, *Soil Microbiology*, Wiley, New York, 356p.

Warming, E., 1909, *Oecology of Plants,* Clarendon Press, Oxford, 422p.

Washburn, A. L., 1973, *Periglacial Processes and Environments*, St. Martins Press, New York, 320p.

Wright, H. E., Jr., 1968, History of the Prairie Peninsula: Quaternary of Illinois, *Illinois Univ. College of Agri. Spec. Pub. 14,* pp. 78-88.

PART
II

TRANSITIONAL DEPOSITS

GRAVITY DEPOSITS

Gravity deposits, mostly colluvium, are of many kinds and are gradational between residuum and transported deposits. Most gravity deposits develop by becoming lubricated with water (or ice), and most occur on steep slopes. They grade from residuum into stream, glacial, eolian, or other transported deposits where the transporting medium—water, ice, or wind—dominates the movement.

The map includes colluvium with untransported deposits but distinguishes some varieties depending on different kinds of bedrock (shown by subscripts). Because of the small scale it does not distinguish the very important related deposits: landslides, avalanches, mudflows, boulder fields, talus heaps, rockfalls, or collapse structures. The map groups all these together as colluvium (co).

COLLUVIUM

Colluvium properly includes only those deposits that are or have been moving slowly downslope by ground creep (Fig. 3-1). In general the deposits contain considerable fine-grained material and support substantial growths of forest or shrub.

Although colluvium is generally a chaotic mixture of coarse- and fine-grained materials, the deposits locally form boulder fields. Some of these represent avalanches of coarse debris; others are lag concentrates of the larger fragments that are left after the fine-grained materials are washed away. At the other extreme, when soaking wet, colluvium grades into mudflows of fine-grained material viscous enough to float or to roll boulders. Thicknesses of colluvium are generally less than 10 feet (3 m) and rarely more than 25 feet (7.6 m), but they may grade into thick debris cones at the base of hillsides. Texture, thickness, and extent depend on parent materials, steepness of slope, and climate.

Land inventories, by both the U.S. Geological Survey and the Forest Service, indicate that in hilly or mountainous humid regions like the Appalachians and the Pacific Northwest, colluvium covers 95 percent or more of the

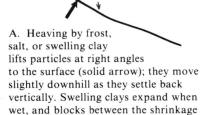

A. Heaving by frost, salt, or swelling clay lifts particles at right angles to the surface (solid arrow); they move slightly downhill as they settle back vertically. Swelling clays expand when wet, and blocks between the shrinkage cracks fill with material washed, blown, or fallen into them.

B. The trajectory of particles thrown into the air by rain splash is longest in the downhill direction.

C. Falling trees uproot the ground. The decaying stump and earth holding the root produce a mound and sediment from uphill becomes rolled or washed into the depression.

D. Burrowing animals remove soil and keep it in a mound below the burrow.

E. Livestock or deer trails along hillsides knock loose materials down slope and form a flat path. These are not to be confused with naturally forming terracettes in patterned ground.

F. Where creep is active, trees may be bent. Bending accompanied by tilting produces a "crazy forest".

G. Creep can tilt telephone or power lines.

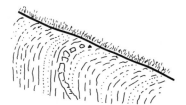

H. Steeply inclined strata or veins become bent downhill in the actively moving surface layer.

Figure 3-1. Cross sections illustrating some processes and effects of creep.

ground. In semiarid mountains, like the Rocky Mountains, the isolated mountains on the Colorado Plateau, and the leeward slopes of the Pacific Coast mountains, colluvium covers 85 to 90 percent of the surface, including the summits where frost heaving produces orderly ground patterns (see Fig. 2-10). In arid regions, like the southern part of the Basin and Range Province, colluvium may cover 70 percent or less of the mountain slopes.

Figure 3-2 illustrates the kind of stratigraphy to be found in colluvial deposits in the Southwest. The youngest deposit occupies present-day gullies and is composed of stones with little or no rock stain. Such deposits may be forming at the present time, or they may be subject to erosion by gully deepening where the debris covers a shale slope. The deposits may extend all the way to the caprock, while downslope they grade into the alluvium of the present floodplain. On either side of the gully containing the modern colluvium, there may be ridges of bedrock capped and protected by older colluvium. Stones on these older deposits are generally stained with iron and manganese oxides, and many grade into alluvium that can be dated archaeologically. In the examples shown in Figure 3-2, the oldest colluvium caps the still higher ridges between the Middle Holocene deposits and grades downslope to Late Pleistocene alluvium. The positions of these deposits provide a measure of the rate of cliff retreat at the location. On the Colorado Plateau, the Great Plains, and the Basin and Range Province, the Middle Holocene colluvium commonly indicates cliff retreat of shale or other weakly consolidated rocks amounting to 5 to 10 feet (1.5 to 3 m); the late Pleistocene colluvium indicates retreat of 25 to 50 (8 to 15 m) feet. These estimates of cliff retreat in those semiarid regions are quite compatible with the shorter range estimates of slope retreat based on bristlecone pine.

In the eastern United States, especially in the northern Appalachians, much of the colluvium is thought to have formed during deglaciation. The colluvium is younger than the glacial deposits and overlies them (Fig. 3-3).

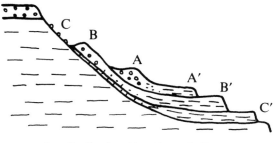
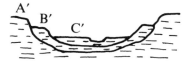

longitudinal section through the hillside.

transverse section through A' showing how the alluvial terraces are inset in each other.

Figure 3-2. Cross section illustrating typical stratigraphy of colluvial deposits, as exposed in the western United States. Three ages of alluvium can be dated archaeologically and by their overlapping relationships: A' (oldest), B', and C'. The youngest colluvium (C) collects in present-day gullies between ridges capped and protected by the next older deposit (B), which in turn is between still higher and older ridges capped and protected by colluvium A. These colluvial deposits grade downward into the datable alluvial deposits.

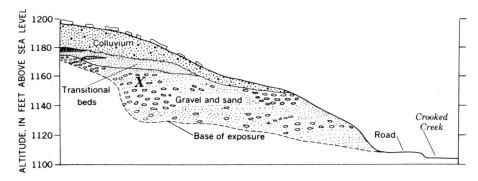

Figure 3-3. Gravity deposit in a glaciated valley. Colluvium overlying stratified glacial drift, gravel, and sand near Tioga, Pennsylvania. At the position marked X, the articulated remains of a Pleistocene pig were recovered. The colluvial slopes are stable at the present time and are thought to have formed soon after the glacial deposits were laid down, while the ice was retreating. (From Denny and Lyford, 1963)

Apparently the preglacial colluvium was swept out by the glaciers and incorporated in the drift. Much of the postglacial colluvium was deposited shortly after the glaciers retreated because the underlying drift shows no signs of weathering and the present rate of mass wasting is too slow to produce the observed thickness of colluvium.

Shale slopes capped by resistant, slabby caprock may become completely armored by downhill creep of slabs of the caprock. Figure 3-4 is an example from the Great Plains in northeastern Colorado.

Some colluvial deposits form terracettes (Fig. 3-5), a form of patterned ground (see Chapter 2). These are not to be confused with hillside livestock trails (Fig. 2-1*E*), which tend to be long, narrow, and continuous whereas the naturally formed terracettes are generally short, lobate, and discontinuous.

GRAVITY DEPOSITS NOT SHOWN ON THE MAP

Landslides, debris avalanches, and mudflows are local deposits too small to show individually on the map, yet they are of major importance because of the many kinds of hazards associated with their unstable ground. The deposits are most common where slopes are steep, where formations include much shale, and where water can infiltrate and lubricate the ground.

Landslides, debris avalanches, rockfalls, and mudflows are related. They differ chiefly in the proportion of water in the earth being moved and consequently the mechanism of movement. Rockfalls may be dry. Most landslides move en masse on a lubricated slip surface. Mudflows move by flowage, and debris avalanches move partly by slippage and partly by flowage. A given mass displaced by gravity may be in segments that move in different ways. Downslope a landslide may become an avalanche, and farther down the avalanche may become a mudflow.

Figure 3-4. Shale slopes capped by slabby sandstone may become completely covered (armored) by fallen slabs, and the armor slows the erosion. This example is on the Great Plains in northeast Colorado.

Landslides

Landslides are those masses of rock that have moved downslope as a unit or complex of units and that are elongated parallel to the contour. They are most common along shale cliffs but may occur on slopes as low as 5 percent (5 feet in 100 feet) (1.5 m in 30.5 m). Most landslides are rotated backwards, creating a poorly drained depression at the top of the slide. These depressions are important in the mechanics of movement of such slides because they collect water that can seep into the base of the slide, further lubricating the plane of sliding. As a consequence, such slides form unstable ground subject to renewed movement. A complex of slides, in plan, forms a crescentic pattern on the hillside (Fig. 3-6).

Some slumped masses break into parts that become rotated downhill. Slumping may also occur where bedding slopes downhill, as in the geologic setting of Rainbow Bridge, in southern Utah. Fears have been expressed that the rise of Lake Powell may sufficiently lubricate the bedding to cause downhill creep and collapse of the bridge.

Much or most of the damage caused by the earthquake at Anchorage, Alaska, in 1964 was the result of sliding of ground along a nearly horizontal bed of clay underlying the glacial deposits that had provided the foundations for buildings. The earthquake shocks caused the clay to liquefy (like puddling quicksand by tapping it) so that the glacial deposits were temporarily floating on a nearly frictionless surface. The resultant sliding produced a series of horsts and grabens that toppled residences, schools, and other buildings on the cracked ground.

Avalanches

Avalanches, or debris avalanches, are caused by rapid flowage of poorly consolidated materials. Snow avalanches in mountains are a familiar example. Debris

Figure 3-5. Terracettes, or steps, in Death Valley evidently date from a wetter time, such as the Middle Holocene, when there was a lake 30 feet (9.14 m) deep in the valley, or the Late Pleistocene, when there were even deeper lakes. But the terracettes are now stable, as indicated by lack of movement dowels over a 4-year period. These discontinuous lobate steps are not due to livestock.

avalanches, like colluvium, are composed of a chaotic mixture of coarse- and fine-grained materials, but colluvium moves slowly, by creep, whereas avalanches move rapidly, by flowage. Avalanches (Fig. 3-7) differ from landslides in several ways. As mentioned, they move by flowage rather than by slippage; they are a chaotic mixture of materials rather than discrete blocks of bedrock; and generally they are narrow deposits elongated downslope rather than short, wide deposits paralleling the hillside contours. Some landslides grade into avalanche deposits in the downslope direction (Fig. 3-8).

Debris avalanches develop where the ground has become so saturated that internal friction within the mass of debris is reduced and it can flow as a viscous mass. Avalanching is common in poorly consolidated deposits that can collect enough moisture to become saturated. For example, snowbanks or hard rains on steep slopes can convert colluvium to an avalanche. Highway cuts that intersect the water table can cause small avalanches.

In the arid and semiarid Southwest, avalanche deposits are numerous, and some evidently date from the glacial or pluvial periods of the Pleistocene when there was more moisture than there is today. In many places the mass of material that moved seems excessive for any single storm, and a possible explanation is that these avalanches represent the accumulation of large snowbanks during several seasons. Great quantities of water could be discharged during a severe thaw. Figure 3-7 illustrates an occurrence on the Colorado Plateau, where present annual precipitation is less than 10 inches (25 cm). However, in high-rainfall areas (>20 inches) present-day avalanching is a common hazard (Figs. 3-8, 3-9).

N

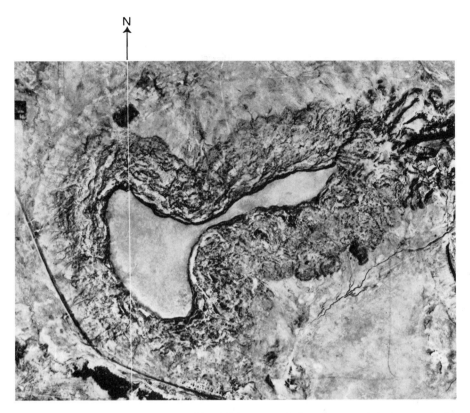

Figure 3-6. Concentric patterns of landslides around a basalt-capped shale mesa in the Mt. Taylor volcanic field, New Mexico. North is at the top. The flat top of the mesa is about 3½ miles (5.6 km) long. The village of Grants is at the lower center.

Unstable ground jarred by an earthquake can create dry avalanches. Where the Madison River crosses the Madison Range in Montana, the Hebgen Lake earthquake in 1959 jarred loose a huge mass of rock that was unstable because its structure dipped into the canyon (Fig. 3-10). Nearly 40 million cubic yards (30.4 million cubic meters) of rock broke from the south wall, flowed across the floor of the canyon, and moved 400 feet up the north wall. When it came to rest it dammed the Madison River, producing what now is known as Earthquake Lake. This tragic avalanche swept across a campground and buried an estimated 25 campers. Such avalanches may move on a layer of air trapped under the debris; the air becomes compressed by the load on it so the debris can flow as a frictionless mass before it settles to the ground. Excavations at some avalanches that moved onto the surfaces of glaciers reveal delicate ice structures preserved under the debris.

Some slides that probably began during the Pleistocene and are still moving are marked by crazy forests of tilted trees (Fig. 3-1*G*). One example is the Slumgullion slide that dammed Lake Fork above Lake City, Colorado. Another example is the crazy tilted forest just north of the Bonneville Dam on the

Figure 3-7. Debris avalanche on the semiarid Colorado Plateau at Frye Canyon, Utah, about 5-8 miles (8-13 km) west of Natural Bridges National Monument. The rim on the skyline is about 7,000 feet (2.134 m) in altitude; the floor of the valley at the debris avalanches is about 5,500 feet (1,676 m). Between the ridges jutting into the canyon near the base of the escarpment are avalanche deposits derived from the high cliffs. The deposits are attributed to sudden thaws of Late Pleistocene snowfields that accumulated on the shady, north-facing slopes of the escarpment. View is to the south.

Figure 3-8. Debris avalanche (long, narrow, light area) in the Wasatch Plateau, Utah (about 17 miles (27.4 km) west of Orangeville). The slide, about 1 mile (1.6 km) long, dammed the big valley into which it moved. Headward, toward the big cliff, the avalanche grades into landslides. Displacement probably began during a wet period when snow collected in the shade of the cliffs, which face north, probably at least as long ago as the Middle Holocene, but there has been renewed movement during subsequent wet periods and as recently as 1971.

Figure 3-9. View of the slide at Thistle, Utah. The slide is old (Late Pleistocene?) but was reactivated by heavy rain and snow in early April 1983. The reactivated slide, 1½ miles (2.4 km) long and ¼ mile wide, overwhelmed the D. & R.G.R.R. and U.S. Highway 6-50. The toe of the slide buried the track and highway under 200 feet (61 m) of debris that formed a dam across the canyon bottom. The resulting lake completely submerged the town of Thistle, which was located just upstream from the toe of the slide. A new railroad tunnel was excavated through the mountain from which this picture was taken, and service was restored in July. The highway was closed for more than 6 months, necessitating a 75-mile (120.7 km) detour for mainline traffic.

Columbia River, where present-day movement has required the moving of power lines that had become tilted.

Mudflows

Mudflows contain more water and are more fluid than avalanches and may spread widely on gentle slopes if not confined (Fig. 3-11). The content of sediment is so great, however, that large, heavy boulders can be floated. As the flow spreads it moves more slowly, and the larger and heavier boulders settle to the ground, further checking the rate of flow. The water continues on as rivulets carrying fine sediment in suspension. After the flow has ceased the ground may be strewn with the coarse debris, much of the fine sediment having been washed away.

Mudflows are common at the mouths of canyons draining easily eroded

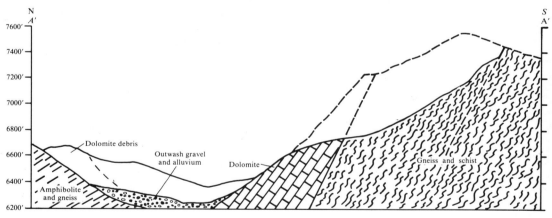

Figure 3-10. View and cross section of the Madison Canyon slide that was caused by the 1959 earthquake at Hebgen Lake, Montana. Nearly 40 million cubic yards of bedrock and colluvium slid off the south wall of the canyon and dammed the Madison River. Northward-dipping foliation and shear zones in the schist and gneiss contributed to the instability and movement of the rock. (Photograph courtesy of Irving Witkind; cross section after Hadley, 1964)

formations. Hard rains or melting snowbanks wash mud and stones into the gullies, and the streams become overloaded and converted to a fluid paste, with too much solid matter to be a stream but with sufficient water to permit flowage. When dried, the debris is added to and becomes a part of the alluvial fan at the canyon mouth. Figure 3-11 illustrates this process; the mudflow originated in badlands of Tertiary shale.

Another sort of mudflow, a relict of the Pleistocene, is illustrated in Figure 3-12. These mudflows originated in granitic rocks high in the Panamint Range on the west side of Death Valley and moved onto the apex of a fan at the foot of

Figure 3-11. Mudflows are fluid enough to spread widely yet pasty enough to transport stones (cf. Fig. 3-13). As flowage slows the stones settle to the ground, and much or most of the fine sediment is carried away as sheetwash in rivulets. This mudflow, in Death Valley, originated in fine-grained Tertiary formations. Another mudflow near this location left a mound 10 feet (3.1 m) high across the highway, which was replaced by new pavement laid across the top of the mound. (Photography by John Stacy, U.S. Geological Survey)

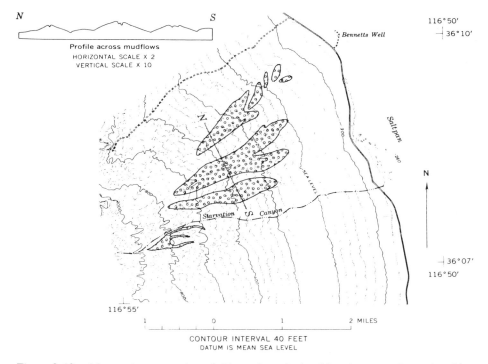

Figure 3-12. Map and cross section of ridges of granite boulders (patterned areas) making relict mudflows derived from granitic intrusions high in the Panamint Range on the west side of Death Valley. Many of the boulders are the size of the one illustrated in Figure 3-13 and have been transported 10 (16.1 km) miles from their source. (From Hunt and Mabey, 1966)

the mountain. The deposits form ridges 2 to 3 miles (3 to 5 km) long, 500 to 1,000 feet (152.4 to 304 m) wide, and 50 to 75 feet (15 to 22 m) high. They include abundant boulders of granite 5 to 10 feet (1.5 to 3 m) in diameter. Each ridge has a narrow crest with a dry wash along it (see cross section Fig. 3-12); the sides are strewn with the boulders and slope to the adjoining fan surface. The volumes of the deposits range from 8 to 25 million cubic yards (6.1 to 19 million cubic meters). Probably they formed in the Pleistocene at a time when snowbanks on the Panamint Range accumulated over a period of several years and could provide the necessary volume of water when there was a sudden thaw.

Somewhat similar deposits have been found at the north foot of desert mountains in Utah (Fig. 3-13). These probably began in the mountains as landslides of diorite porphyry sliding on Cretaceous shale, but became mudflows downslope as the blocks of rock broke and churned together. If there was continued seepage in the shale under the deposit, there may have been renewed movement over a long period of time. These deposits are probably also due in part to abrupt thawing of accumulated Pleistocene snowbanks high in the mountains that provided the large quantity of water needed to move such large boulders.

A third kind of deposit common on steep mountainsides occurs where torrential runoff causes flash flooding that can move coarse stones and leave them as a natural levee along each side of the gully (Fig. 3-14).

Figure 3-13. Relicts of Pleistocene mudflows form scattered boulders at the foot of many desert mountains. This example is at the north foot of the Henry Mountains, Utah. Those mountains were not glaciated, but during the Pleistocene pluvial periods, perennial snowfields evidently formed on the north slopes, and sudden thaws could create mudflows capable of transporting boulders like these 10 miles from their source. These particular boulders are strongly pitted as if by solution (cf. Fig. 2-8).

Boulder Fields and Related Periglacial Features

Boulder fields are numerous in mountainous areas where resistant rock formations are or have been subject to severe frost action (Fig. 3-15). They are present far south of the limit of glaciation and at altitudes below those reached by alpine glaciers. They are also found in cirques and other areas that were glaciated; these boulders are younger than the glacial features and partly mask them. For example, steep cirque walls commonly have boulders collected at or near their bases. They were broken from the cirque walls by the intensive frost action that continued after the glacier had melted. In some cirques the boulders continue to accumulate.

Most boulder fields are stable and show no evidence of downhill creep at the present time (but see Fig. 3-1*F*). Some may have moved to their present position by sliding while being held together by ice; others may have formed as avalanches; still others seem to have formed by slow downhill creep. At places this creep takes the form of episodic sliding of the individual boulders past one another.

Many boulder fields have running water under the boulders long after snow on the ground surface has melted. Snow falling on the boulder field sifts into the empty spaces between the boulders where, because of shade from the boulders, spring and summer melting is slower than on the surrounding soil-covered ground. Water, discharged from under the boulders, emerges at spring zones at the lower ends of the boulder fields. These are marked by distinctive,

Figure 3-14. Natural levees of cobbles are thrown out at each side of gullies subject to mudflows caused by flash flooding. Examples are widespread on the western mountains. This channel on the Panamint Range, California, is about 10 feet (3.1 m) wide, and its levees are 1–2 feet (0.3 to 0.61 m) high.

Figure 3-15. Boulder fields develop wherever resistant rock formations have been subjected to severe frost action. They are extensive on the tops of glaciated mountains but also occur on high mountains and plateaus that have been subject to severe frost action although not glaciated. The Henry Mountains, for example, were not glaciated, and their summits are a mass of loose boulders that have been shifted but not far out of place. Taking a pack string across such ground can be exciting.

water-loving vegetation that contrasts with the surrounding and drier mountain side.

Ground Subsidence, Collapse Structures

Many or most of Florida's lakes are the result of the collapse of cavernous limestone; some areas of seemingly stable ground have collapsed under the weight of residences built on cavernous ground. Where salt beds are thick, as along the lower stretches of the Pecos River in southeastern New Mexico and western Texas, solution of the salt has caused the collapse of overlying formations. The sinks containing ponded sediment mixed with collapsed blocks (karst topography) are also extensive on limestone or gypsiferous formations throughout the humid part of the country (Fig. 3-16), even on the limestone slopes of the Sacramento Mountains in New Mexico and Kaibab Plateau forming the north rim of the Grand Canyon.

Land subsidence and collapse also occur in clayey deposits by a process known as piping. As water enters the ground, it moves along cracks in the easily eroded clay or shale, and the cracks are widened by erosion. When the cracks are sufficiently wide, the ground sags. Along the banks of arroyos, piping so weakens the banks that they collapse. Some of the prehistoric buildings on the floodplain at Chaco Canyon National Park have been threatened by such collapse. Many arroyos (Fig. 7-4) are widened by piping as much as by lateral erosion by the stream. Piping also causes pitting like karst topography in shale formations containing high percentages of salts. There are good examples in Death Valley near the Visitors' Center.

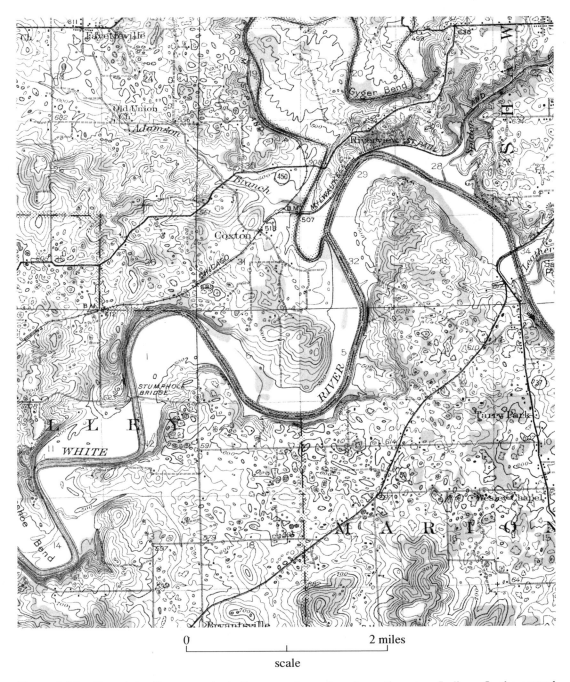

Figure 3-16. Karst (pitted) topography in limestone formations in southwestern Indiana. In the central and eastern United States, cavernous limestone formations are characterized by numerous sinks, where water enters the ground. Drainage therefore is disintegrated; there are few through-flowing streams. Also, though not illustrated here, some streams become "lost rivers" — they go underground at hills and emerge on the other side. (Map from U.S. Geological Survey Oolitic 15-minute quadrangle, 1942)

Similar subsidence may develop on alluvial fans that are composed of much silt and clay. The subsidence has commonly been attributed to settling of the clays, but probably some is caused by piping. Cities on such fans may be damaged, as at Cedar City, Utah. Other fans with subsidence features are found along the west base of the Sierra Nevada.

Ground subsidence accompanied by cracking of the ground and even of buildings may be caused by excessive withdrawals of groundwater or of oil (Fig. 3-17). There are many examples of such "groundwater mining" beside those illustrated, including those in Houston, Texas, Baton Rouge, Louisiana, and the Phoenix area, Arizona. Similar subsidence at Long Beach, California, has been caused by withdrawal of oil. Ground subsidence is common in coalfields, too, due to the underground mining (Fig. 3-18).

A third cause of ground subsidence is the placing of heavy loads on plastic or uncompacted material such as clay beds, peats, or buried organic soils. The Grand Opera House in Mexico City and the Leaning Tower of Pisa are familiar examples. The Washington Monument has settled about 6 inches (152.4 mm)

Figure 3-17.   An example of ground subsidence caused by excessive withdrawal of groundwater. This example is at Las Vegas, Nevada (Photograph by Thos. L. Holzer, U.S. Geological Survey)

because of compaction of a clay bed about 100 feet (30.5 m) below the base of the monument. Its settling is even because the load is distributed evenly and the compactible layer is homogeneous and of even thickness, but a structure may become tilted if the load is irregularly distributed or the compactible layer is irregular.

REFERENCES CITED AND ADDITIONAL BIBLIOGRAPHY

Alden, W. C., 1928, Landslide and flood at Gros Ventre, Wyoming, *Am. Inst. Mining and Metall. Eng. Trans.* **76:**347-361.

Blackwelder, E., 1928, Mudflow as a geologic agent in semiarid mountains, *Geol. Soc. America Bull.* **39:**465-480; discussion by J. T. Singewald, Jr., pp. 480-483.

Cleaves, A. B., 1961, Landslide investigations, *U.S. Dept. Commerce*, 67p.

Denny, C. S., 1961, Landslides east of the Funeral Mountains, near Death Valley Junction, California, *U.S. Geol. Survey Prof. Paper 424-D*, pp. D85-D89.

Denny, C. S., and W. H. Lyford, 1963, Surficial geology and soils of Elmira-Williamsport region, New York and Pennsylvania: *U.S. Geol. Survey Prof. Paper 379*, 60p.

Fryxell, F. M., and L. Horberg, 1943, Alpine mudflows in Grand Teton National Park, Wyoming, *Geol. Soc. America Bull.* **54:**457-472.

Gilluly, J., and U. S. Grant, 1949, Subsidence in the Long Beach Harbor Area, California, *Geol. Soc. America Bull.* **60:**461-529.

Hadley, J. B., 1964, Landslides and related phenomena accompanying the Hebgen Lake earthquake of August 17, 1959, *U.S. Geol. Survey Prof. Paper 435-Km*, pp. 107-138.

Highway Research Board, 1958, *Landslides and engineering practice*, Spec. Rept. No. 29, Washington, D.C.

LaMarche, V. C., Jr., 1968, Rates of slope degradation as determined from botanical evidence, White Mountains, California: *U.S. Geol., Survey Prof. Paper 352-I*, pp. 341-377.

Figure 3-18. Ground subsidence due to collapse of underground coal mines, near Scranton, Pennsylvania.

Morton, D. M., and R. Streitz, 1967, Landslides, *California Div. Mines and Geology Mineral Inf. Service, 20,* (10, 11).

Parker, G. G., and E. A. Jenne, 1967, Structural failure of western highways caused by piping, in Symposium on subsurface drainage, 46th Ann. Mtg., Highway Research Rec. no. 203, Natl. Research Co-Natl. Acad. Sci.-Natl. Acad. Eng. Pub. 1534), pp. 57-76.

Pestrong, R., 1976, Landslides—the descent of man, *California Geology,* July, pp. 147-158.

Poland, J. F., and J. H. Green, 1962, Subsidence in the Santa Clara Valley—A progress report. *U.S. Geol. Survey Water-Supply Paper 1619-C,* 16p.

Richmond, G. M., 1962, Quaternary stratigraphy of the La Sal Mountains, Utah, *U.S. Geol. Survey Prof. Paper 324,* 135p.

Schumm, S. A., and R. J. Chorley, 1966, Talus weathering and scarp recession in the Colorado Plateaus, Ann. Geomorph. 1966, pp. 11-36.

Sharp, R. P., 1942, Mudflow levees, *Jour. Geomorph.* **5:**222-237.

Sharpe, C. F., 1968, *Landslides and Related Phenomena,* Cooper Square Publishers, New York, 137p. (Originally published by Columbia University Press, 1938).

Shroder, J. F., 1971, Landslides of Utah, *Utah Geol. and Mineralog. Survey Bull. 90,* 51p.

Shroder, J. F., and R. E. Sewall, 1985, Mass movement in the La Sal Mountains, Utah, in Contributions to Quaternary geology of the Colorado Plateau, G. E. Christenson et al., eds., *Utah Geol. and Mineral Survey Spec. Studies 64,* pp. 49-85.

Taber, S., 1929, Frost heaving, *Jour. Geology* **37:**428-461.

Terzaghi, K., and R. B. Peck, 1967, *Soil mechanics in engineering practice,* John Wiley, New York, 729p.

Troll, C., 1958, Structure, soils, solifluction, and frost climates of the earth, *U.S. Army Corps of Engineers Snow and Permafrost Res. Estab., Translation 43,* 121p.

Washburn, A. L., 1973, *Periglacial processes and environments,* St. Martins Press, New York, 320p.

Williams, G. P., and H. P. Guy, 1971, Debris avalanches, a geomorphic hazard; in *Environmental geomorphology,* D. R. Coates, ed., SUNY, Binghamton, New York, pp. 25-46.

Woodford, A. O., and T. F. Harriss, 1928, Geology of Blackhawk Canyon, San Bernadino Mountains, California, *Calif. Publ. Geol. Sciences Bull.* **17:**265-304.

Wooley, R. R., 1946, Cloudburst floods in Utah, *U.S. Geol. Survey Water-Supply Paper 994,* 120p.

PART
III

TRANSPORTED
DEPOSITS

Most surficial deposits are sediments, weathered from bedrock in one area, transported by water, wind, ice, or gravity, and deposited in another area, where they await further weathering erosion and removal. Thus, they are not only very much younger than the underlying bedrock, they also are usually unrelated to it. Residual deposits, described in Chapter 1, are exceptional. As already noted (Chapter 3), gravity deposits are intermediate; they are not residually in place, but neither are they far out of place. On the map, transported deposits are classified according to the mode of transportation by which they were moved: shore deposits (Chapter 4); glacial deposits (Chapter 5); stream deposits (Chapter 6); lake deposits (Chapter 7); and eolian (wind) deposits (Chapter 8).

SHORE DEPOSITS

COASTAL DIMENSIONS AND TYPES

Both the length of the U.S. coast and the thicknesses of surficial deposits along it depend on scale and definitions. On the 1/7,500,000 map the coasts measure approximately as follows: Gulf Coast, 2,000 miles (3,218 km); Atlantic Coast, 2,500 miles (4,023 km); Pacific Coast (California to Washington), 2,500 miles (4,023 km). However, if these distances are measured on larger-scale maps, the measurements are very much larger. For example, because of the many embayments along the Atlantic Coast, its measured length increases from about 2,500 to 28,500 miles (4,023 to 45, 870 km) when the scale of measuring is increased from 1/7,500,000 to about 1/500,000. In contrast, the straighter Pacific Coast increases only about 3½ times.

The practical importance of our shores is illustrated by the fact that more than half the country's population lives in the coastal counties, and the percentage is increasing. By the time this book is published, the percentage may be 75 percent (National Map Atlas, 1970, p. 55). Land management problems will increase.

Rocky cliffs constitute about 25 percent of the Atlantic Coast and about 90 percent of the Pacific Coast of California, Oregon, and Washington. About half of each of these stretches of cliff is rock; the other half consists of sand, some of it gravelly, in bayhead and bayside beaches inset in the cliffs and in baymouth bars and spits (Fig. 1-III A). Cliffs in poorly consolidated rocks that erode easily are subject to landsliding, a notable real-estate hazard along much of the California coast and along comparatively short segments of the Atlantic Coast, such as the Atlantic Highlands in New Jersey and the Calvert Cliffs in Chesapeake Bay. Erosion is minimal along rocky coasts like that of Maine (Fig. 4-1) and the Puget Trough in Washington (Fig. 4-2). The Gulf Coast is without cliffs except for low limestone bluffs along the west coast of Florida.

A second kind of shore, sandy shores (shown on the map by the hachured coast line), forms barrier beaches and/or sandy plains back of estuaries (Fig. 4-3). These shore deposits represent about 75 percent of the Atlantic Coast

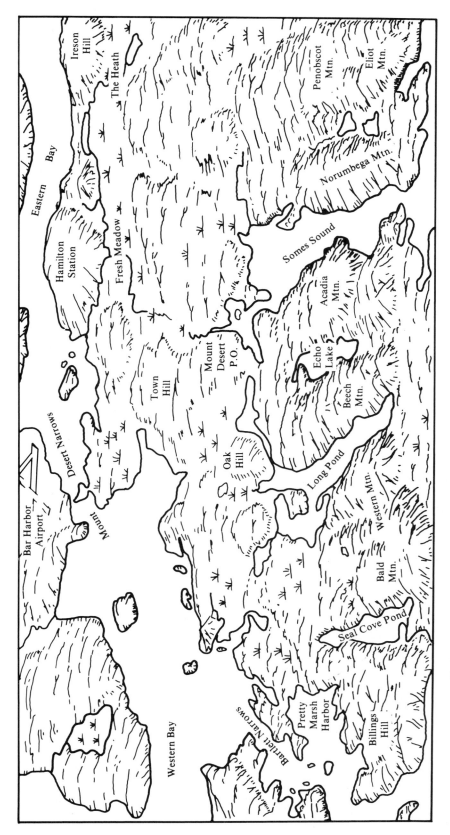

Figure 4-1. Diagrammatic view landward (north) of Bar Harbor area, Maine, illustrating the irregular, rocky, embayed coast of New England. Valleys have been deepened by glacial scour; hills are round. Ponds and marshes dot the till-covered lowlands. Width of view is about 10 miles (16.1 km).

Figure 4-2. Diagrammatic view of the Olympic Peninsula and Puget Sound, Washington. Width of view is about 75 miles (121 km). Unlike the Pacific Coast farther south, Juan de Fuca Strait and the trough at Puget Sound were glaciated. Alpine glaciers on the Olympic Peninsula did not reach the sea, but their meltwaters contributed outwash to the shores.

including the sea islands (Si on the map) of Georgia and South Carolina, 60 percent of the Gulf Coast, but only about 10 percent of the Pacific Coast, other than the narrow beaches along or inset with the cliffs. The problem of measurement is illustrated by Padre Island on the Texas coast and the estuary, Laguna Madre, behind it; each is about 100 miles (161 km) long and together they have 300 miles (483 km) of sandy beach!

A third major kind of ground along the coasts consists of wetlands, which are marshy mudflats of various kinds (bm on map). Much of it is deltaic, as along the delta of the Mississippi River. Some tidal flats are perennially submerged; others are submerged only at high tide. Each area of wetland has its distinctive vegetation depending on muddiness, degree and frequency of submergence, salinity of the water, and latitude. The most extensive wetlands are characterized by grasses and rushes in estuaries back of beach ridges. These are widespread along both the Gulf and Atlantic coasts, as shown on the 1/7,500,000 map. They are of limited extent along the Pacific Coast, but because wetlands are the most productive of our biotic zones, those limited Pacific Coast stretches are of special ecological importance. A very different kind of wetland is formed of mangrove swamp on coral sand around the south tip of Florida (cr on the map). Swampy ground back from the Gulf and Atlantic shores includes cypress swamps (as far north as Maryland) and hardwood forest in poorly drained bottoms of the Dismal Swamp and similar bottoms in the southeastern states.

The Mississippi River delta (Fig. 4-4, see also Fig. 2-1), shown as al and bm on the map, is a unique part of the U.S. coast. It is the youngest part of the United States coast and is still being extended seaward. The ground at New Orleans was not built above sea level until about the time of the Battle of Hastings (A.D. 1066). Fringing the delta and paralleling the coast are ridges of sand and shell a few feet high, commonly supporting stands of evergreen oak, known as chenieres. These are relicts of former barrier beaches separated by poorly drained bottomlands. Other features of the Gulf Coast include hummocky

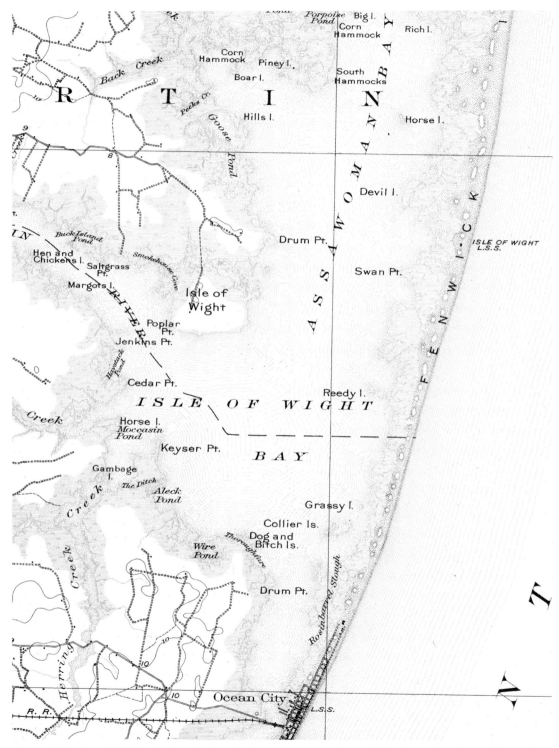

Figure 4-3. Fenwick Island, a barrier beach along the eastern shore of Maryland, has dammed the valleys and formed bays. (U.S. Geological Survey 1/62,500 quadrangle map)

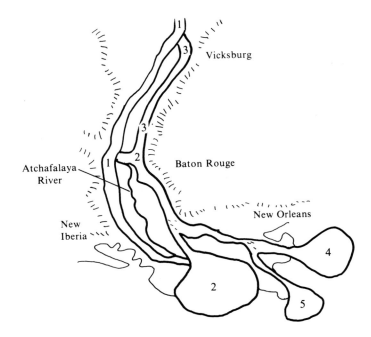

Figure 4-4. Five stages in the growth of the Mississippi River delta. 1: Pre-A.D. 1. 2: Diversion about A.D. 300-400. 3: Diversion about A.D. 1000-1100. 4: Diversion about A.D. 1100-1200. 5: Diversion to present channel about A.D. 1500-1600. (After Fisk and McFarlan, 1955)

ground (not shown on the map) at salt domes. The highest, Avery Island (8 miles [13 km] south of New Iberia, about 100 miles [161 km] west of New Orleans) is almost 200 feet (61 m) higher than the surrounding marshes. Five such mounds form the Five Islands along the shore of the Mississippi River delta from New Iberia to the Atchafalaya. Other kinds of less extensive shore deposits include hummocky, bouldery, and gravelly till, as along the north shore of Long Island, New York. Locally along some beaches, gravels or even rubbly beach litter may become cemented by calcium carbonate to form beachrock.

All our coasts have harbors that are sites of highly developed shipping and other industrial uses. Although percentagewise they are of minor extent, they are the sites of maximum environmental disturbance by construction, dredging, and dumping of wastes offshore.

EUSTATIC AND OTHER CHANGES IN SEA LEVEL

In addition to the variable effects of weathering and erosion, a major factor contributing to differences in coast lines has been the eustatic changes in sea level. Withdrawal of water from the oceans lowered sea level during the glaciations, and there was a corresponding rise when the ice melted during the interglaciations.

The problem of sea-level changes, however, is complicated by two kinds of earth movement. The continent sank isostatically when it became loaded with glacial ice, and it rebounded again when that load of ice was removed by melting during the interglaciations, including the present, the Holocene. In addition, there have been tectonic movements, rather obvious along the earthquake prone Pacific Coast, but there has also been warping along the Atlantic Coast. Finally, there is the further complication of land subsidence as a result of compaction of mud at places like the Mississippi River delta and ground subsidence due to withdrawal of groundwater (e.g., Galveston area, Texas; San Jose Valley, California) and of oil (e.g., Long Beach, California).

At the time of the last glacial maximum (ca 25,000 years ago), sea level was a few hundred feet lower—perhaps 500 feet (152 m)—than it is now. Numerous elephant remains have been dredged from ground that is now submerged as part of the continental shelf. Well-preserved stumps of eastern white pine, dated by radiocarbon at 4,000 years B.P. were found in the intertidal zone on the Maine coast. The stumps indicate eustatic rise of sea level of at least 9 feet (2.7 m) in 4,000 years. The fish weir discovered in an excavation at Boston indicates a rise of 16 feet (5 m) in 4,500 years. At Prince Edward Island, Canada, intertidal peat and radiocarbon-dated tree stumps indicate 5 to 8 feet (1.5 to 2.4 m) of submergence during the last 900 years.

Submerged deposits of peat and of intertidal shellfish have been dredged from numerous places on the continental shelf off the mid-Atlantic and New England states. The dated samples, from depths as great as 65 meters (210 feet), provide a generalized curve for the rise of the sea level during the Holocene. The data suggest that the sea level was 400 feet (121 m) below its present position 15,000 radiocarbon years ago, rose rapidly to about 20 meters (65 feet) below its present level about 7,000 years ago, and has risen slowly since. Measurements differ, but the present rise along that part of the coast probably averages about 15 cm (ca. 6 inches) per century.

The continent has been tilted northeastward, with the result that the coastal plain gradually slopes below the sea off the south coast of New England. This tilting probably occurred during the middle or late Tertiary. The northeast part of the continent was further depressed by the former load of glacial ice and has only partly rebounded as the load of ice has melted. However, in northern New England and along the St. Lawrence lowland, the isostatic rise of the continent exceeded the rise of the sea level. Along the coast of Maine, marine beach deposits are 200 feet (61 m) above sea level. Other earth movements along the seemingly stable Atlantic Coast include downwarping in South Carolina and Georgia, which has enabled the sea to advance inland, producing the sandy sea islands (si on the map). Some of this advance is recent enough to have killed oak trees. These displacements, from Maine to South Carolina, are considerable, even by California standards.

The Pacific Coast, including a part in Oregon and southern Washington that is seismically quiet, is characterized by uplifted marine terraces that record substantial amounts of earth movement in late Pleistocene times. There is present-day movement along the seismically active stretches of that coast. The faults are not shown on the map.

GULF COAST DEPOSITS

Shore deposits along the Texas coast have been eroded from deltaic or alluvial fan formations. Inland the deposits form a series of irregularly palmate branching, broad sandy ridges traversing areas that are predominantly clay. These Pleistocene deposits consist of a lower sand (Lissie Formation) about 500 feet thick, including some gravels and containing remains of *Elephas*, *Equus*, and *Bison*. The formation has been interpreted as a glacial-stage deposit formed when the sea level was lower. Near the coast it is overlain by deltaic clay and sand (Beaumont Clay), also about 500 feet thick and containing marine beds interbedded with marsh and swamp deposits, including cypress logs. The Holocene shore deposits along the eroded edge of the Pleistocene formations include sand at the barrier beaches, sand in dunes back of the estuaries, and alluvium in the valleys (Fig. 4-5).

Off the coast of Louisiana, west of the mouth of the Mississippi River, are Quaternary deposits 8,000 feet (2,438 m) thick and apparently entirely of marine origin. These beds have been correlated with 2,000 feet (610 m) of alluvial deposits near New Orleans (Akers and Holck, 1957). Figure 4-6 illustrates the kind of stratigraphy; the shifting positions of the marine and freshwa-

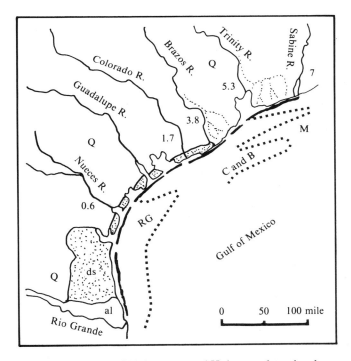

Figure 4-5. Map of Pleistocene and Holocene deposits along the Texas coast. Average annual discharge of rivers is indicated in millions of acre-feet. Along most of the coast are barrier beaches (black). These and the Holocene sand along the shore of the estuaries overlap Pleistocene marine clay and deltaic deposits. On the floor of the Gulf of Mexico (out to about 60 fathoms) the river sediments are drifting northward (*RG*, Rio Grande sediment; *C, B*, Colorado and Brazos rivers sediment; *M*, Mississippi River sediment). Trinity River sediment is forming a delta at the head of the bay. Pleistocene courses of the Trinity and Brazos rivers are shown by dotted lines. (Based on data from numerous sources).

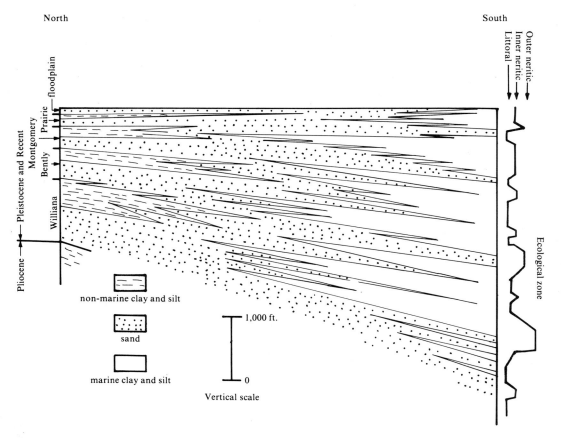

North   South

Figure 4-6. Diagrammatic section illustrating intertonguing of marine deposits (white) and continental deposits (stippled) in the Mississippi River delta in southern Louisiana. The ecological zones indicated in the marine deposits were inferred from faunal assemblages obtained in drill cuttings. Correlations chiefly by electric well logs. Length of section about 100 miles (161 km). (After Akers and Holck, 1957)

ter deposits may reflect changes in depth of water due to floods, compaction of muds, and/or eustatic changes in sea level. Figure 2-1 is a view of the marshy ground (bm on the map) forming the surface of much of the Mississippi River delta.

The general geology of the Gulf Coast east of the Mississippi River delta is similar to that of the Texas coast. Pleistocene deposits west of the Apalachicola River consist of two superimposed sequences of terrigenous clastic sediments (Schnabel and Goodell, 1968). Each sequence becomes finer grained upward. The upper part is interpreted to represent a mid-Wisconsinan transgression of the sea, and the lower is interpreted to represent a Sangamon transgression. Two shorelines in the Apalachicola River area are the Pamlico, 25 to 30 feet (7.6 to 9 m) above sea level, and the Silver Bluff, 5 to 10 feet (1.5 to 3 m) above sea level. A Sangamon age is inferred for the Pamlico, correlating it with the uppermost part of the Beaumont Clay in Texas, and a mid-Wisconsinan age is inferred for the Silver Bluff.

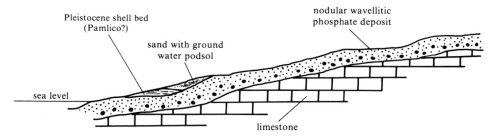

Figure 4-7. Pleistocene shell bed (Pamlico?) overlain by sand with groundwater Podzol and overlying nodular, wavelletic phosphate deposits. West coast of Florida, area of Tampa Bay.

The thickest Holocene deposits along the eastern Gulf Coast are found in the incised valley of the Apalachicola River, which was eroded deeply during the last lowering of the sea level. Deltaic and bay sediments filled the valley as the sea level rose during the Holocene.

Although Florida is a peninsula composed largely of limestone, much of its surface is sand that has a shallow water table. At the Everglades are extensive marsh deposits with peat; elsewhere on the peninsula are numerous cypress marshes (Fig. 2-2). The Gulf Coast shore is partly sand in barrier beaches, backed by estuaries with marshy ground (bm), partly cliffs cut into the limestones, and, towards the south, partly coral (cr) and partly mangrove swamp. On the west coast, near Tampa, a phosphate deposit (equivalent to saprolite and composed of the mineral wavellite) can be traced westward under a Pleistocene shell bed (Fig. 4-7).

ATLANTIC COAST DEPOSITS

Most of the shore from Florida to Cape Cod is characterized by sand (s), with short stretches of cliffs (not shown separately on the map) at places like the Calvert Cliffs in Chesapeake Bay and Navesink Highlands in New Jersey just south of Sandy Hook (Fig. 4-11). The sandy shorelines are of different kinds, and each has distinctive shapes, as shown on the 1/7,500,000 map.

In Florida the sand beaches are wide, straight, and miles in length. The Georgia and South Carolina coasts are marked by sea islands (si) (Fig. 4-8), many with firm sand beaches and some with dunes; much of the ground between the islands is salt marsh. The sea islands coincide with a downwarp, and the islands are probably the result of eustatic rise of sea level coupled with downwarping. To the north, from near Charleston to Cape Hatteras, is a stretch with cuspate sandy beaches, some with narrow barrier beaches backed by lagoons (Figs. 4-9, 4-10). This stretch coincides with a structural arch, the Cape Fear Arch, an uplift 2,500 feet (762 m) higher structurally than the sea islands downwarp. Next north is the embayed section of the coastal plain where the formations dip northeast off the Cape Fear Arch and the river valleys have been embayed all the way to the line of falls, the Fall Line, at the edge of the Piedmont Plateau.

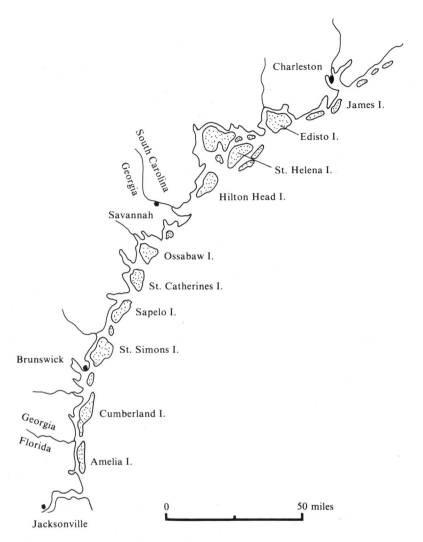

Figure 4-8. Sea islands along the Georgia–South Carolina coast are the result of the sea advancing into the downwarp; locally the encroachment is recent enough that the stumps of killed oak trees are still standing in the salt water. This local coastal downwarping, coupled with gradual eustatic (worldwide) rise of sea level, results in a remarkable relative sea-level rise in this area.

The stratigraphy is confused. In the Charleston area, red and yellow soil developed on Pleistocene marine deposits is attributed to pre-Wisconsinan weathering. The slight weathering of the Pamlico terrace suggests an age no greater than mid-Wisconsinan. In southern Virginia, deposits along the Suffolk scarp (base 25 feet [7.6 m] above sea level; equal to the Pamlico?) have been interpreted as Sangamon.

In parts of Florida, peat deposits containing a cold-water flora of diatoms

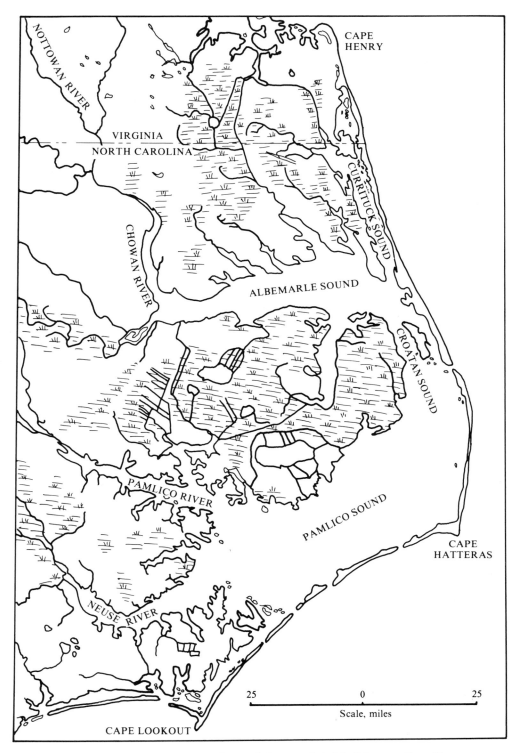

Figure 4-9. Map of barrier beaches and the sounds behind them at Cape Hatteras.

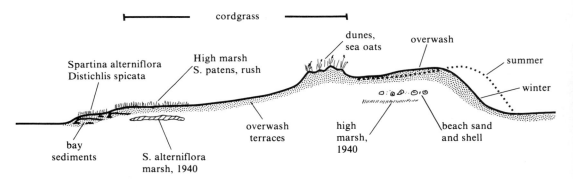

Figure 4-10. Cross section of the Cape Hatteras barrier reef as interpreted by Godfrey and Godfrey (in Coates, 1973).

overlie a brackish water deposit (Pamlico?). In Georgia, Pleistocene vertebrates overlie a shell bed that possibly is the Pamlico. In South Carolina Pleistocene fossil vertebrates are associated with the Pamlico. In southern Virginia and North Carolina the Pamlico is overlain by 10 feet of peat (Richards, 1936). In Maryland and New Jersey marine deposits at about the 25-foot (7.6 m) contour have been correlated with the Pamlico. At all these latitudes, the fauna of the Pamlico indicates an environment as warm as or warmer than that of the present, and it is inferred that the deposit formed during an interglaciation. However, there is question about the correlations because of possible warping.

Curious but still unexplained features of the mid-Atlantic coast are the Carolina bays, which are oval depressions (meteorite scars?), most with ponds or marshes a few feet to more than 30 feet (9 m) deep and ranging from a few to hundreds of acres. Most are oriented northwest-southeast. Rims are sandy, especially on the southeast side. Although best known in the Carolinas, they occur as far north as Maryland.

Upland gravels (shown as saprolitic residuum, rg on the map) in Virginia and Maryland help reconstruct the history of the drowned and embayed valleys. The gravels are alluvial fan deposits in which the rivers became deeply incised when sea level was lowered during the glaciations. At the mouth of the Susquehanna River, terrace gravels 200 feet (61 m) higher than the river are weathered to saprolite (Fig. 1-6) and are evidently pre-Wisconsinan.

The deposits are undergoing continual change. Besides those caused by eustatic rise of sea level, changes are caused by hurricanes and other storms. Figure 4-11 shows progressive changes in the shoreline at Sandy Hook as recorded by successive surveys.

Surficial deposits on Long Island include the sandy beaches and bars along the south shore and two end moraines known as the Ronkonkoma Moraine on the south and Harbor Hill Moraine at the north. Under these deposits are older tills and Pleistocene marine clay, all resting on Tertiary coastal plain formations. This geology has been a major factor in the occurrence, development, and pollution of groundwater on Long Island.

The moraines extend eastward from Long Island to Martha's Vineyard and Cape Cod (Figs. 4-12, 4-13). Beyond there they are submerged, and the coast north of Cape Cod is made of rocky cliffs with sand only at bayside and bayhead

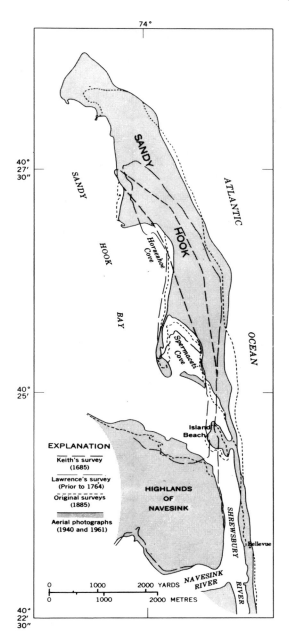

Figure 4-11. Map of spit at Sandy Hook, New Jersey, showing progressive changes in the shoreline during the period 1685-1961. (From Minard, 1974)

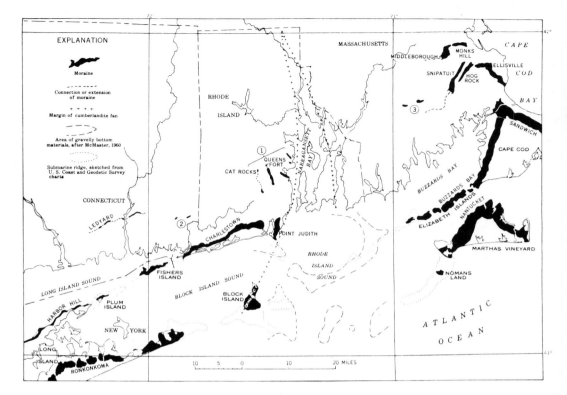

Figure 4-12. End moraines and submerged gravel ridges east of Long Island, New York. (From Schafer, 1961)

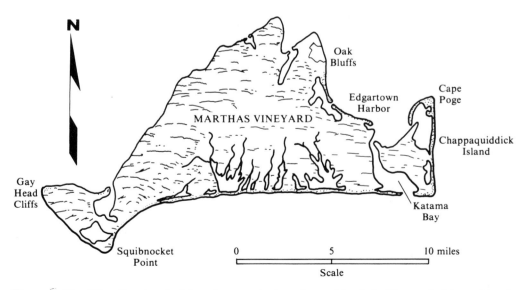

Figure 4-13. Shoreline in glacial and outwash deposits at Martha's Vineyard, Massachusetts. Wisconsinan end moraines form the hilly northwest and northeast sides of the island. Sandy baymouth bars have closed the mouths of valleys draining the outwash plain south of the moraine.

bars. North of Boston isostatic rise of the land has exceeded the eustatic rise of the sea level, and the coastal marine deposits consist of sand, silt, and clay that grade landward into deltaic deposits, which in turn grade up the valleys to glacial outwash. The limit of the marine deposits rises northward from 15 meters just north of Boston to 60 to 90 meters in southwest Maine and 120 meters in northern Maine (Chute and Nichols, 1941; Upson and Spencer, 1964; Bloom, 1960).

COASTAL TERRACES SOUTH OF LONG ISLAND

Sandy deposits on terraces inland from the sandy shores of the Atlantic coast have been interpreted as marine in origin (Cooke, 1945; MacNeil, 1950). Their heights, the names commonly assigned to them, and their supposed ages are (Cooke, 1943, 1945): 270 feet (82.3 m), Brandywine, Aftonian; 215 feet (65.5 m), Coharie, Yarmouth; 170 feet (51.8 m), Sunderland; 100 feet (30.5 m), Wicomico (= Surry), Sangamon; 70 feet (21.3 m), Penholoway; 42 feet (12.8 m), Talbot; 25 feet (7.6 m), Pamlico (=22 Suffolk?); 5 feet (1.5 m), Silver Bluff. However, there is increasing evidence that the terraces, except the lowest ones, are of fluviatile origin (Carlston, 1950; Hack, 1955), like those in Texas.

CONTINENTAL SHELF DEPOSITS

Surficial continental shelf deposits (Fig. 4-14) are becoming of increasing interest because of uncertainty about the toxic effects caused by pollution of the bottom sediments, a problem discussed in the last section of this chapter.

PACIFIC COAST

The Pacific Coast, the leading edge of the westward-drifting continent, differs markedly from the Atlantic Coast, the trailing edge, and the differences show abundantly in the surficial geology and related geomorphic shore features. The 1/7,500,000 map shows that the Pacific Coast, unlike the Atlantic Coast, consists of alternating short stretches of sandy shores and rocky shores. Beaches are mostly narrow, rocky strips at the foot of wave-cut cliffs, and the widest — but still narrow compared to those along the Atlantic and Gulf coasts — are bars across the mouths of a few streams. Sea cliffs with rocky beaches are most common. Some of these are wave-cut bases of mountains that reach the sea; many are the cliffs at the eroded edges of Pleistocene marine terraces. Many cliffed terraces are a few hundred feet above sea level, and a few are more than 1,000 feet (304.8 m) high, the result of active upfaulting along the coast. Except in northernmost California, Oregon, and southern Washington, the coast is seismically active compared to the Atlantic Coast. Displacements due to earth movements are conspicuous along the Pacific Coast, and many of the uplifted terraces are extensive enough to record warping and tilting. Sand dunes, like the beaches, are spotty, although in places they form a belt wide enough to have migrated inland over forest. Finally, the continental shelf off the Pacific Coast is

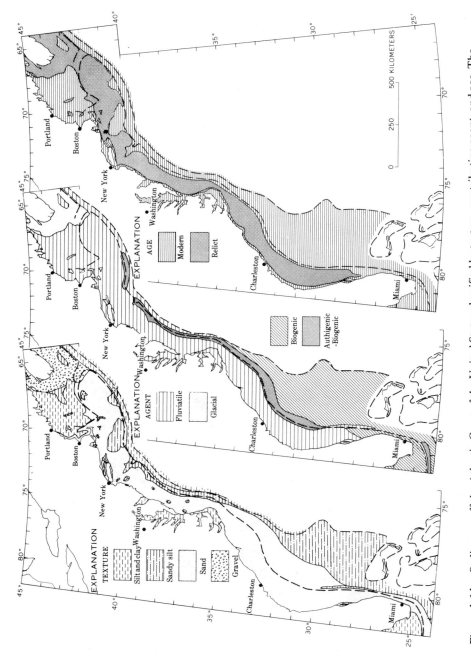

Figure 4-14. Sediments off the Atlantic Coast of the United States classified by texture, contributing agent, and age. The charts are based on inspection of about 1,000 samples prior to detailed laboratory analyses. (From Emery, 1966, Fig. 6)

muddy and narrow and slopes steeply seaward as compared to the broad, gently sloping, sandy shelf off the Atlantic Coast. Figures 4-15 and 4-16 give some details about the shore deposits, shelf, and related geomorphic features along the California and Oregon coasts. South of Point Conception and the Santa Barbara Channel (Fig. 4-15) is an unique area of shelf, suggestive of a submerged part of the Basin and Range Province. Block mountains form islands and are separated by downfaulted or downfolded troughs.

Marine terraces west of Los Angeles in the Ventura region and at the Palos Verdes Hills are similar. In the Ventura region they reach a maximum altitude of 700 feet (213 m); at the Palos Verdes Hills the highest are above 1,000 feet (305 m). The terraces are mostly less than 1,000 feet (305 m) wide; the marine sand and gravel capping them is generally less than 10 feet (3 m) thick. Fossils in the terrace deposits are almost entirely forms living off the southern California coast. The terraces are Late Pleistocene and truncate steeply dipping Early Pleistocene siltstone, sandstone, and conglomerate. Mid-Pleistocene deformation in California was considerable and in much of the state the surficial deposits have been involved. The sequence of events inferred by Putnam (1954, p. 45) is: (1) Deposition of several thousand feet of Early Pleistocene marine and nonmarine sediments in gradually sinking basins such as the Ventura and Los Angeles Basins; (2) folding and faulting of these beds; (3) erosion to reduce the region to moderate relief; (4) pulsatory uplift of individual folds and/or fault blocks and cutting of marine terraces into them; (5) folding of some of the terraces.

Marine faunas off the California coast can be divided into zones that reflect temperature differences due to latitude and, more important, water depth. In the Pleistocene deposits, some species that are locally extinct now live farther north; others live farther south. The faunas of Early Pleistocene formations (Timms Point Silt and Santa Barbara Formation), for example, are found in cool water, whereas the Late Pleistocene (Palos Verdes Sand) is found in warm water. These differences could reflect glacial and interglacial stages, but detailed studies indicate there is little basis for correlating these deposits with particular glacial or interglacial stages (see, for example, Woodring, Bramlette, and Kew, 1946; Valentine, 1961). Correlations have been frustrated because of several difficulties, such as: comparing equivalent faunal facies where earth movements along that seismically active coast have affected water depths; distinguishing faunas that include dead shallow-water shells mixed with living ones in deep water; evaluating changes in outline of the coast, whether due to erosion, sedimentation, or earth movement; and local changes in ocean temperature caused by episodic changes in oceanic circulation.

The cliffed parts of California's coast are being eroded rapidly, both in the geologic sense and, in places, in the sense of real estate damage. Much of the coast is highly subject to landsliding.

From Point Conception north to Monterey Bay the coast is cliffed; sand beaches are narrow and discontinuous, except for an alluvial plain at the mouth of the Santa Ynez River (Fig. 4-15). Sandy deposits on the shelf give way to mud a few miles offshore. Monterey Bay has a submarine canyon that may have been eroded when (and if) structural warping caused San Francisco Bay to discharge southward instead of westward via the Golden Gate. However, the extensive deposit of sand opposite the Golden Gate (Fig. 4-15) suggests consid-

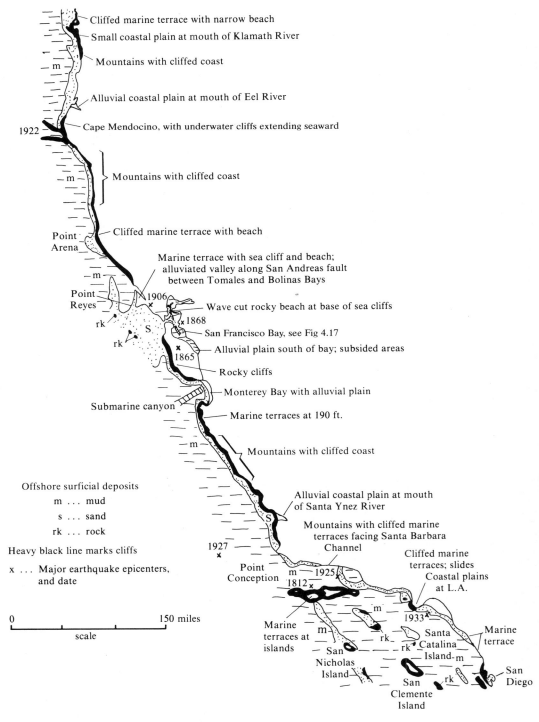

Figure 4-15. Surficial deposits on the continental shelf off the coast of California (after Welday and Williams, 1975), and some of the surficial deposits and geomorphic features along the shore. Black is cliffed coast. Compare with Fig. 4-14.

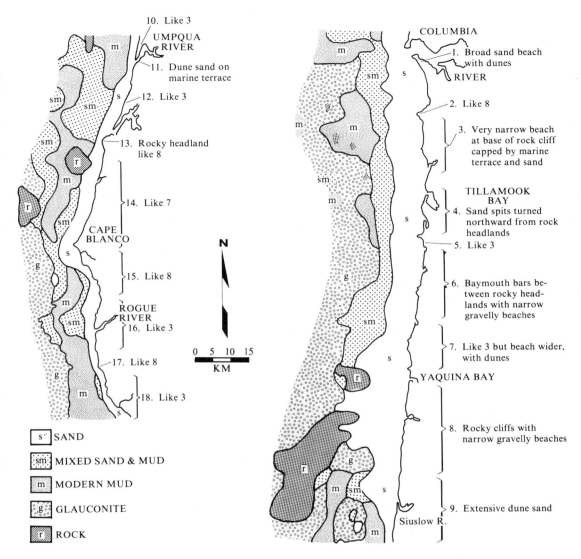

Figure 4-16. Surficial deposits of the Oregon continental shelf, and some of the surficial deposits and geomorphic features along the shore (data mostly from Lund 1971-1975).

erable antiquity for that outlet. The submarine canyon opposite Monterey Bay may have begun as a stream eroded channel, in part or wholly submarine. Apparently it is being maintained and perhaps even deepened by present day submarine currents.

Northward the annual precipitation increases, and north of Monterey Bay the vegetation is dense and the ground is weathered more deeply. This northward increase in precipitation is reflected, too, in the increased number of major rivers.

The San Andreas and other coastal faults extend northwest through Point Reyes and Point Arena to Cape Mendocino, where they turn west into the

Pacific, evidently forming the submarine cliffs west of the Cape (Fig. 4-15). South of Cape Mendocino the coastal area is seismically active; north of there it is quiescent. Curiously, though, along the Oregon coast, uplifted marine terraces are numerous, extensive, and hundreds of feet high. Some are indicated on Figure 4-16. The surficial geology indicates considerable Pleistocene earth movements, but the seismographs indicate little present-day activity from Cape Mendocino to Puget Trough.

The coast of Oregon, about 310 miles (499 km) long (Fig. 4-16) has been described in several papers (see especially Lund, 1971-1975). At the south is a rugged mountainous shore where the Pacific washes against the base of the Klamath Mountains. This stretch is characterized by rocky promontories, sea stacks, rocky clefts in the cliffs, natural bridges, and arches. Farther north much of the coast is benchland of uplifted marine terraces. The cliffed sides of the terraces overlooking the ocean are mostly bordered by narrow beaches (less than ¼-mile [0.4 km] wide) that slope steeply seaward; a few are wider than 1½ miles (2.4 km). Locally, back of the cliffs is a marine terrace capped by sand and surmounted by knobs that are vestiges of former sea stacks. Landsliding is common where the bedrock under a terrace or on a mountainside consists of Tertiary sedimentary rocks rather than igneous or metamorphic rocks.

There is a seasonal change in longshore drift between summer, when the winds are from the northwest and the shore drift is southward, and winter, when the winds are from the southwest and the shore drift is northward. Many of the streams have small flow during dry seasons and may be completely blocked by sand bars. Both summer and winter winds have a landward component so the belt of dunes has encroached landward. Introduction of European beachgrass has stabilized many of the dunes.

The Pacific coast of Washington is somewhat like that of Oregon, but Juan de Fuca Strait and Puget Trough are rocky fjords that are wholly unlike any other parts of the coasts that have been described.

ENVIRONMENTAL CONSIDERATIONS OF THE COASTS

The kinds of problems that arise in trying to manage our coasts are only partly technological; chiefly they are social, economic, and political. The technological problems include:

1. Controlling erosion and sedimentation, especially the changes caused by hurricanes, tsunamis, and storms associated with high tides.
2. Maintaining inputs of fresh water relative to the flow of tidal salt water.
3. Preventing or minimizing the inflow of fertilizers and other nutrients that cause eutrophication.
4. Finding disposal sites for municipal wastes and spoil from dredging sites.
5. Eliminating inflows of biologically harmful chemicals.
6. Improving sewage plants.
7. Coping with changes in circulation caused by deepening of channels, construction of canals, building of groins, sea walls, and so forth.
8. Developing efficient, ready means of cleaning up oil and other chemicals dumped or spilled into bays and onto beaches.

The political, social, and economic problems arise over the need for multiple use of our shores. There are conflicts between the needs for shipping and navigation; protection against the impact of large storms; swimming, boating, fishing, and other kinds of recreation; commercial fishing; preservation of natural areas; and waterfront residential and property owners. The technological problems are easy compared to the kind that require reconciling the conflicts between actual and potential multiple uses.

San Francisco Bay is a notable example of the problems. A tenth of the bay has been filled and about half the remainder is potentially fillable. As the saying goes, "You'll wonder where the water went, when you fill the bay with sediment." The fill has reduced the water surface by a tenth and has greatly reduced the wetlands. Moreover, the fill is an earthquake hazard because when it is shaken it is subject to sliding, a condition like that at Anchorage, Alaska, where there was a disastrous sliding during the 1964 earthquake (p. 62). Such problems at international transportation hubs are not local.

REFERENCES CITED AND ADDITIONAL BIBLIOGRAPHY

Akers, W. H., and A. J. Holck, 1957, Pleistocene beds near the southern edge of the continental shelf, southeastern Louisiana, *Geol. Soc. America Bull.* **68:**983-991.

American Association of Petroleum Geologists, 1972, *Continental shelves origin and significance*, Selected papers reprinted from the Association Bulletin, 1946-1970, Reprint Series No. 3, Tulsa, Okla.

American Association of Petroleum Geologists, 1974, *East Coast Offshore Symposium*, v. 586, June 1974, p. 1055-1239.

Bakker, E., 1972, An Island Called California, University of California Press, Berkeley, Calif., 361p.

Barton, D. C., 1930, Deltaic coastal plain of southeast Texas, *Geol. Soc. America Bull.* **41:**359-382.

Bascom, W., 1964, *Waves and Beaches,* Doubleday (Anchor books) New York.

Bloom, A. L., 1960, *Late Pleistocene changes in sea level in southwestern Maine, Maine Geol. Survey,* Augusta, 143p.

Bloom, A. L., and M. Stuiver, 1963, Submergence of the Connecticut Coast, *Science* **139:**332-334.

Bradley, W. C., and G. B. Griggs, 1976, Form, genesis, and deformation of central California wave-cut platforms, *Geol. Soc. America Bull.* **87:**433-449.

California Division of Mines, 1951, San Francisco Bay Counties Guidebook, *Calif. Div. Mines Bull. 154.*

California Division of Mines, 1954, Geology of southern California, *Calif. Div. Mines Bull. 170.*

Carlston, W. W., 1950, Pleistocene history of coastal Alabama, *Geol. Soc. America Bull.* **61:**1119-1130.

Carson, R., 1955, *The Edge of the Sea,* Houghton Mifflin, New York.

Carter, C. H., 1976, Lake Erie shore erosion, Lake County, Ohio: Setting, processes, and recession rates from 1876 to 1973, *Ohio Geol. Survey, Rept. of Inves. 99.*

Chute, N. E., and R. L. Nichols, 1941, The geology of the coast of northeastern Massachusetts, *Mass. Dept. Public Works and U.S. Geol. Survey Coop. Geol. Proj. Bull. 7,* 48p.

Coates, D. R., ed., 1973, *Coastal Geomorphology,* Publications in Geomorphology, State University of New York, Binghamton, New York.

Collias, E. E., N. McGary, and C. A. Barnes, 1974, *Atlas of physical and chemical properties of Puget Sound and its approaches,* University of Washington Press, Seattle, Wash., 235p.

Cooke, C. W., 1943, Geology of the coastal plain of Georgia, *U.S. Geol. Survey Bull. 941,* 121p.

Crandell, D. R., 1964, Pleistocene geology and piedmont glaciers of the southwestern Olympic Peninsula, Washington, *U.S. Geol. Survey Prof. Paper 500-B,* pp. 135-139.

Crandell, D. R., 1965, The glacial history of western Washington and Oregon, in, *The Quaternary of the United States,* H. E. Wright, Jr., and D. G. Frey, eds. Princeton University, Princeton, N.J., p. 341-353.

Doering, J. A., 1956, Review of Quaternary surface formations of Gulf Coast region, *Am. Assoc. Petroleum Geologists Bull.* **40:**1816-1862.

Doering, J. A., 1960, Quaternary surface formations of southern part of Atlantic coastal plain, *Jour. Geology* **68:**182-202.

Easterbrook, D. J., 1969, Pleistocene chronology of the Puget Lowland and San Juan Islands, Washington, *Geol. Soc. America Bull.* **80:**2273-2286.

Emery, K. O., 1966, Atlantic continental shelf and slope of the United States, geologic background, *U.S. Geol. Survey Prof. Paper 529-A,* 22p.

Emery, K. O., and others, 1967, Freshwater peat on the continental shelf, *Science* **158:**1301-1307.

Emery, K. O., and L. E. Garrison, 1967, Sea levels 7,000 to 20,000 years ago, *Science* **157:**684-687.

Fisk, H. N., and E. McFarlan, Jr., 1955, Late Quaternary deltaic deposits of the Mississippi River, *Geol. Soc. America Spec. Paper 62,* p. 279-302.

Franke, O. L., and N. E. McClymonds, 1972, Summary of the hydrologic situation on Long Island, New York, as a guide to water-management alternatives, *U.S. Geol. Survey Prof. Paper 627-F,* 59p.

Godfrey, P. J., and M. M. Godfrey, 1973, Comparisons of ecological and geomorphic interactions between altered and unaltered barrier island systems in North Carolina, in *Coastal Geomorphology,* Coates, D. R. ed., State University of New York, Binghamton, New York, p. 239-258.

Goldwaite, J. W., 1971, The St. Lawrence lowland, a previously unpublished manuscript in Pleistocene Geology of the St. Lawrence Lowland, Gadd, N. R., ed., *Geol. Survey Canada Mem. 359,* pp. 114-149.

Hack, J. T., 1955, Geology of the Brandywine area and origin of the upland of southern Maryland, *U.S. Geol. Survey Prof. Paper 267-A,* 43p.

Hauge, C. J., 1972, Tides, currents and waves, *California Geology,* July, p. 147.

Heezen, B. C., 1956, The origin of submarine canyons, *Sci. American,* August, 7p.

Hunt, C. B., 1974, Natural regions of North America, W. H. Freeman and Co., New York, 725p.

Hunt, C. B., and A. P. Hunt, 1957, Stratigraphy and archeology of some Florida soils, *Geol. Soc. America Bull.* **68:**Fig. 3.

Johnson, F., and H. M. Raup, 1947, *Grassy Island,* Phillips Academy, Andover, Mass., Papers of the Peobody Foundation for Archeology, 68p.

Kennedy, M. P., 1973, Sea-cliff erosion at Sunset Cliffs, San Diego, *California Geology,* February.

King, C. A. M., 1972, Beaches and Coasts, 2nd ed. Edward Arnold, London, 570p.

Kurz, H., 1942, Florida dunes and scrub, vegetation and geology, *Florida Geol. Survey Bull. 23,* 154p.

Kurz, H., and K. Wagner, 1957, Tidal marshes of the Gulf and Atlantic Coasts of Northern Florida and Charleston, South Carolina, *Florida State Univ. Studies, No. 24,* 168p.

Lankford, R. R., and J. J. W. Rogers, 1969, *Holocene Geology of the Galveston Bay Area,* Houston Geological Society, Houston, Tex. 141p.

LeGrand, H. E., 1961, Summary of the geology of the Atlantic Coastal Plain, *Am. Assoc. Petroleum Geologists Bull.* **45:**1557-1571.

Lippson, A. J. ed., 1973, *The Chesapeake Bay, Maryland—An Atlas of Natural Resources,* Johns Hopkins University Press, Baltimore, Md., 55p.

Lund, E. H., 1971-1975, Landforms along the coast of Oregon, *The Ore Bin,* **33** (5, 11); **35** (5, 12); **36** (5, 11); **37.**

McKee, B., 1972, *Cascadia—The Geologic Evolution of the Pacific Northwest,* McGraw-Hill, New York, 394p.

Milliman, J. D., and Emery, K. O., 1968, Sea levels during the past 35,000 years, *Science* **162:**1121-1123.

Minard, J. P., 1974, Slump blocks in the Atlantic Highlands of New Jersey, *U.S. Geol. Survey Prof. Paper 898,* 24p.

Muller, E. H., 1965, Quaternary Geology of New York, in *The Quaternary of the United States,* H. E. Wright, Jr., and D. G. Frey, eds., Princeton University Press, Princeton, N.J., p. 99-112.

Murray, G. E., 1961, *Geology of the Atlantic and Gulf Provinces of North America,* Harper and Bros., New York, 692p.

The National Atlas, 1970, U.S. Geological Survey, Washington, D.C., 417p.

Norris, R. M., 1968, Sea cliff retreat near Santa Barbara, California, *California Geology,* June.

Oakeshott, G. B., 1971, *California's Changing Landscape—A Guide to the Geology of the State,* McGraw-Hill, New York, 388p.

Pestrang, R., 1972, San Francisco Bay tidelands, *California Geology,* February.

Powers, C. F., and A. Robertson, 1966, The aging Great Lakes, *Sci. American* **215**(5): 94-104.

Prouty, W. F., 1952, Carolina Bays and their origin, *Geol. Soc. America Bull.* **63:**157-224.

Puri, H. S., and R. O. Vernon, 1964, Summary of the geology of Florida, *Florida Geol. Survey Spec. Pub. 5,* 312p.

Putnam, W. C., 1954, Marine terraces of the Ventura region and Santa Monica Mountains, California, in R. H. Jahns, ed., Geology of Southern California, *California Dept. Nat. Res. Div. Mines Bull. 170,* p. 45-48.

Rau, W. W., 1973, Geology of the Washington coast between Point Grenville and the Hoh River, *Wash. Dept. Nat. Resources Bull.* 66, 58p.

Richards, H. G., 1936, Fauna of the Pleistocene Pamlico formation in the southern Atlantic Coastal Plain, *Geol. Soc. America Bull.* **47:**1611-1656.

Richards, H. G., and S. Judson, 1965, The Atlantic Coastal Plain and the Appalachian Highlands in the Quaternary, in H. E. Wright, Jr., and D. G. Frey, eds., *The Quaternary of the United States,* Princeton University Press, Princeton, N.J., pp. 129-136.

Schafer, J. P., 1961, Correlation of end moraines in southern Rhode Island, art. 318, *U.S. Geol. Survey Prof. Paper 424-D,* pp. D68-D70.

Schnabel, J. E., and H. G. Goodell, 1968, Pleistocene-Recent stratigraphy evolution, and development of the Apachicola coast, Florida, *Geol. Soc. America Spec. Paper 112,* 72p.

Shepard, F. P., 1956, Late Pleistocene and Recent history of the central Texas coast, *Jour. Geol.* **64:**56-69.

Snavely, P. D., Jr., and H. C. Wagner, 1963, Tertiary geologic history of western Oregon and Washington, *Washington Div. Mines and Geology Inv. 22.*

Snavely, P. D., Jr., and N. S. MacLead, 1971, Visitor's guide to the geology of the coastal area near Beverly Beach State Park, Oregon, *The Ore Bin* **33** (5) p. 85-108.

Sorensen, R. M., 1968, Recession of marine terraces—with special reference to the coastal area north of Santa Cruz, Calif., in *American Society Civil Engineers, 11th Coastal Engineering Conf. Proc.* **1:**653-670.

Starke, G. W., and A. D. Howard, 1968, Polygenetic origin of Monterey submarine canyon, *Geol. Soc. America Bull.* **79:**813-826.

Sullivan, R., 1975, Geological hazards along the coast south of San Francisco, *California Geology*, February.

Trumbull, J., and others, 1958, An introduction to the geology and mineral resources of the continental shelves of the Americas, *U.S. Geol. Survey Bull. 1067,* 92p.

Upson, J. E., and W. W. Spencer, 1964, Bedrock valleys of the New England coast as related to fluctuations of sea level, *U.S. Geol. Survey Prof. Paper 454-M,* 44p.

Valentine, J. W., 1961, Paleoecologic molluscan geography of the Californian Pleistocene, *Calif. Univ. Dept. Geol. Sci. Bull.* **40:**309-442.

Van Andel, T. H., and D. M. Poole, 1960, Sources of recent sediments in the northern Gulf of Mexico, *Jour. Sed. Petrology* **30:**91-122.

Vedder, J. G., and others, 1969, Geologic framework of the Santa Barbara Channel region, in Geology, Petroleum Development, and Seismicity of the Santa Barbara Channel Region, *U.S. Geol. Survey Prof. Paper 679,* pp. 1-11.

Welday, E. E., and J. W. Williams, 1975, Offshore surficial geology of California, *Calif. Div. Mines and Geol., Map Sheet 26.*

Whitmore, F. C., Jr., and others, 1967, Elephant teeth from the Atlantic continental shelf, *Science* **156:**1477-1481.

Woodring, W. P., 1952, Pliocene-Pleistocene boundary in California Coast Ranges, *Am. Jour. Sci.* **250:**401-410.

Woodring, W. P., M. N. Bramlette, and W. S. W. Kew, 1946, Geology and paleontology of Palos Verdes Hills, Calif., *U.S. Geol. Survey Prof. Paper 207,* 145p.

GLACIAL DEPOSITS

Glacial deposits cover most of North America north of Long Island and north of the Ohio and Missouri rivers—nearly 20 percent of the conterminus United States. They are readily divisible into an older and a younger series of deposits on the basis of stratigraphy and differences in the degree of weathering and erosion. The younger deposits, those of the Wisconsinan glaciation (Table 1), are almost as little weathered as when they were deposited by the glacial ice and its meltwaters. Moreover, they are only slightly eroded and still preserve their constructional landforms. The older glacial deposits, formed during the pre-Wisconsinan glaciations, are as strongly weathered and considerably dissected by erosion as saprolite. Their landforms, therefore, are of erosional rather than depositional forms. Moreover, in the central United States they are largely covered by eolian silt (loess).

Glaciers require not only cold but moisture to develop; during the Pleistocene there was plenty of both. Glaciers are of two principal kinds: continental ice sheets, illustrated today by the icecaps at Greenland and Antarctica; and mountain, or alpine glaciers (mg on the map), including icecaps at high-altitude plateaus. The most southerly alpine Pleistocene glaciers in the United States were located at the San Bernadino Mountains, California, San Francisco Peak and White Mountains, Arizona, and a very small one, too small to show on the map, at Sierra Blanca in the Sacramento Mountains, New Mexico. Today some perennial snowfields and small glaciers cover about 200 square miles (518 square km) in mountainous parts of the western states. The most southerly of these are in the Sierra Nevada and the mountains of Colorado.

During the early part of the Wisconsinan glaciation, the ice sheets, which developed over Canada, had two principal source areas: one the Labradoran Ice Sheet, centered in Labrador and eastern Quebec, and the other, the Keewatin Ice Sheet, centered on the plains of Manitoba and the adjacent parts of Ontario and Hudson Bay. Later, another center, the Patrician, located south of Hudson Bay, sent ice lobes southward through the Lake Michigan basin and westward and southwestward through the Lake Superior basin into northern Wisconsin and northeastern Minnesota. The Keewatin lobes were crowded westward but received sufficent snowfall and had enough energy to reach as far south as Des Moines, Iowa. Over the mountains of western Canada a complex of coalescing glaciers were molded into valleys between peaks (nunataks) that projected above the ice. These western deposits (tg on the map) extended southward across the lowland bordering Puget Trough and extended across Juan de Fuca Strait onto the north foot of the Olympic Peninsula (Fig. 4-2).

Deposits of the Wisconsinan glaciation vary greatly in composition from one area to another depending on the topography covered by the ice, differences in kind of bedrock from which the deposits were derived, and rates of ice advance and retreat. New England hills and mountains, for example, are hard crystalline rocks that were the source of large and small boulders, or glacial erratics (Fig. 5-1), in the drift (tg) of that area. Glacial erratics are widely strewn over New England, and the boulders cleared from fields were used to build the stone walls so characteristic of that part of the country. Many are striated and some are polished (Fig. 5-2). The granite boulders on the north shore of Long Island were carried from New England by the glacial ice and deposited on the poorly consolidated sandstone and shale of the coastal plain formations. Similarly, in New York, boulders of Precambrian rocks from the Adirondack Mountains were transported southward and deposited on Paleozoic formations; some of them were carried onto the Catskill Mountains.

One of our better-known glacial erratics is Plymouth Rock. Another, located near Leyden, Massachusetts, about 15 miles (24 km) from its probable source, measures 10 feet (3 m) long, 6 feet (1.8 m) wide, and 5½ feet (1.7 m) high; although it weighs 20 tons, it is so well balanced that it can be swayed by hand.

In the central United States, the glacial drift contains few boulders (compare Figs. 5-3, 5-4); the source areas, except, notably, the Superior Upland, did not have mountains to supply them. The Precambrian crystalline rocks eroded there were transported southward and deposited on Paleozoic sedimentary

Figure 5-1. Granitic glacial erratic near Greensboro, Vermont. (Photograph by Wayne Newell)

Figure 5-2. Striae and polish on a glacial boulder. The striae are at right angles to the foliation (dark grooves with a knife along one of them). A few shiny patches of glacial polish are preserved just above the left end of the knife. Location near St. Johnsbury, Vermont.

Figure 5-3. Glacial till in New England characteristically has many boulders. Location near Danville, Vermont.

Figure 5-4. Glacial till in the Middle West characteristically has few boulders, as here at Iron River, Wisconsin.

formations in the Central Lowland. Pipestone National Monument, Minnesota, has a cluster of large erratics of granite rocks similar to those in New England. In Montana, southward advance of the glacial ice across the Missouri Plateau is recorded by stones plucked from porphyritic rocks in laccolithic intrusions (Sweetgrass Hills, Bearpaw Mountains, Little Rocky Mountains) and deposited in trains extending southward from the mountains. Such trains show the direction of ice advance; so, too, do the glacial striae.

Causes of the climatic changes that led to glaciations have been widely discussed for more than a century by geologists, astronomers, meteorologists, and physicists, and we still have no satisfactory explanation. Among the hypotheses that have been considered are: changes in the amount of volcanic or cosmic dust in the atmosphere; changes in cloudiness; polar wandering or shifts in positions of the continents; structural displacements on the continents or ocean floors causing shifts in ocean and air currents; eccentricity of the earth's orbit around the sun; and variations in solar radiation.

The total volume of glacial ice existing today is still not accurately known, and the total volume during the glacial maxima is even more uncertain. The quantity during glacial maxima may have approximated a layer of water about 100 to 150 meters thick over the oceans. A dozen estimates suggest that the volume of present ice equals a layer of water 50 to 60 meters thick over the oceans.

Geologists divide the Quaternary period into the Pleistocene and Holocene (Recent) epochs (Table 1). The Tertiary period preceding the Quaternary was a comparatively warm period; the Pleistocene epoch of the Quaternary period was alternately cold and warm, and the cold periods were times of glaciations. Ice accumulated in northern Canada at least four times in amounts sufficient to form continental masses that pushed southward into the United States. Similar and apparently correlative ice sheets formed in northern Europe and spread southward. In Europe, however, five major glaciations have been recognized, and the earliest may predate the earliest glaciation recognized in this country. In addition to the continental ice sheets, there are alpine, or high-altitude glacial deposits (mg on the map) in the mountain valleys in the western United States and Canada.

Deposits closely associated with the glacial drift but described separately in this report include alluvial deposits far removed from the drift but attributable to glacial outwash, notably along the Mississippi River (Chapter 7), lake deposits (Chapter 6), peat deposits (Chapter 2), eolian (wind-blown) silt (loess) derived in large part from glacial outwash (Chapter 8), gravity deposits (periglacial boulder fields, landslides, avalanches, and colluvium; Chapter 3), and some shore deposits (Chapter 4).

The glacial deposits record progressive overlap eastward of younger drifts over older ones. The Kansan overlaps (extends onto) the east side of the Nebraskan, the Illinoian overlaps the east side of the Kansan, the Early Wisconsinan overlaps the east side of the Illinoian, and, assuming Denny's (1956) correlation of moraines on the Atlantic seaboard, the Late Wisconsinan overlaps the east side of the Early Wisconsinan. Such distribution of maximum extent of the deposits suggests an eastward shifting of the North American ice center or centers during Pleistocene times.

GLACIAL DEPOSITS IN THE NORTHEASTERN UNITED STATES

Facies

Glacial and related deposits in the northeastern states include terminal or end moraines (see map explanation for symbols) that form ridges extending the length of Long Island and eastward to Nantucket and Cape Cod (Fig. 4-12). Westward, at Staten Island and farther west across New Jersey and northern Pennsylvania, the two moraines merge and become one. Farther west, in northern Pennsylvania and New York, at least two end moraines are recognized north of the Valley Heads Moraine (refers to the Finger Lakes valleys), but in New England there are only short, very local moraines. In New England the ice melted by a process referred to as stagnation (Flint, 1930). Instead of the ice front melting back along a broad front, segments in valleys stagnated and became isolated from neighboring ice masses. The result is a seemingly chaotic array of closely associated glacial, fluvio-glacial, and lacustrine deposits of quite different kinds.

However, detailed mapping of the deposits, which are important for all kinds of environmental and engineering considerations, has produced a good deal of order out of the seeming chaos. At any given moment in time, four very different kinds of glacial and glacially related deposits were laid down. At the north, where ice was still continuous, till was deposited; unsorted gravel and sand dropped to the ground as the ice melted. This deposit included materials at the base of the ice (subglacial till), materials within the ice (englacial till), and materials that had collected on the surface of the ice (supraglacial till). The subglacial till in general consists of materials found not far from their source, generally less than 2 miles (3.2 km) away. Englacial till contains materials from farther up the valley, and supraglacial till includes much of the loose material, including big erratics that slid from the hillsides onto the surface of the ice. Thus, even the till has an orderliness when looked at in detail.

Not only is the glacial drift in New England characteristically stony (Fig. 5-3), so too are the modern soils developed on the drift. The composition of the drift and soil varies greatly depending on the bedrock overridden by the ice. Along the Connecticut River valley, for example, the drift is reddened with debris from Triassic redbeds, while on the bordering uplands of crystalline rocks east and west of the valley the drift has gray tones.

Another well-displayed feature of the drift in New England is the occurrence of boulder trains extending more or less southward from hills having distinctive kinds of rocks. These trains, like the glacial striae, show the direction of ice movement. One of the first discoveries of the boulder trains occurred in 1842, when Stephen Reed described a "chain of erratic serpentine rocks" extending several tens of kilometers southeastward into Massachusetts from an amphibolite hill in New York (Holmes, 1966).

At many places glacial striae on bedrock show two directions of ice movement, and these have been interpreted to record two glacial advances. Another interpretation, however, is that the later striae simply record a late stage in the movement of a single ice sheet, when the ice had thinned and changed its direction of movement by adjusting to the topography.

South of the belt of continuous ice was a belt where the valleys contained stagnating, melting blocks of ice bordered by streams and lakes. The stream and lake deposits in this belt are ice-contact deposits, debris deposited in part against walls of ice. Such deposits, formed against walls of ice, collapsed when the ice wall melted. This belt includes kames, kame terraces, deltaic deposits, and bouldery stream deposits. There also are morainelike deposits known as flow till that are attributed to mudflows.

Next south of the belt with stagnating ice is a belt where torrential glacial outwash built coarse fluvio-glacial deposits, and still farther south is a belt where finer-grained fluvio-glacial deposits formed. This fourfold sequence of facies, first recognized by Jahns (1951, 1953, 1966), is graded to a common level, such as a bedrock spillway, and as a consequence, as the sequences progressed up-valley, the younger deposits became inset in the older ones. The facies overlap upstream. At any given place, a fourfold stratigraphic sequence may be recognized with till at the base and fine-grained fluvio-glacial deposits at the top.

In detail, these facies are complexly intertongued. An example is provided in Cambridge, Massachusetts, where the following chronology of events occurred (Chute, 1959, p. 195):

1. Advance of the ice and deposition of ground moraine;
2. Retreat of the ice front and deposition of outwash on the moraine;
3. Deposition of sediments in water ponded on the outwash;
4. Partial readvance of the ice and deposition of more moraine and outwash on the lake sediments;
5. Renewed retreat of ice and covering of earlier deposits by outwash;
6. Second episode of deposition in water ponded on the outwash;

7. Deposition of more outwash;
8. Erosion of valleys in that outwash;
9. Melting of stagnating blocks of ice to form ponds;
10. Deposition of peat and Holocene sand and silt in low places.

In the belt formerly covered by ice, thin deposits of ground moraine form an almost continuous cover over everything except the steepest bedrock slopes. As noted above, however, in the valleys these deposits are covered by later facies of the glacial sequence. Also, in places the moraine has been entirely removed by streams. In general, the ground moraine tends to fill depressions in the bedrock surface and produce a smoother, more gently rolling topography.

Drumlins (parallel black ellipses on the map)

Other ice deposits are drumlins (Figs. 5-5, 5-6), teardrop-shaped hills of unsorted sand and gravel elongated parallel to the direction of ice movement. Some are a half-mile long and 150 to 200 feet (46 to 61 m) high. They consist of tough, compact till having distinct pressure partings parallel to the drumlin surface. Drumlins generally occur in groups; solitary ones are rare. They are common along the Connecticut River valley, and a large group at Boston includes one famous in U.S. history, Bunker Hill. Just how they form remains unclear. Some, but not all, have a bedrock core, and these could be attributed to deposition of subglacial till around the obstacle. The bedrock cores are smoothly abraded on the sides and on the end against which the ice advanced. The lee side is plucked and has a rough angular surface. There has been no satisfactory explanation of the drumlins that do not have bedrock cores.

Figure 5-5. Drumlin next to the New York Thruway, southeast of Rochester.

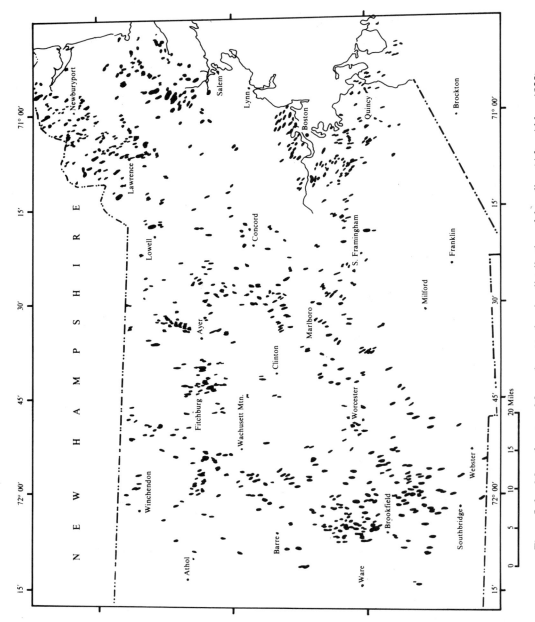

Figure 5-6. Map of eastern Massachusetts showing distribution of drumlins. (After Alden, 1925)

Glacial Stream Deposits (included in tg and ts on the map)

Stream deposits associated with glaciers are of many kinds. Among them are knobby hills of poorly to well-stratified sand and gravel, known as kames (Fig. 5-7). They are 10 to 100 feet (3.1 to 30.5 m) high and occur singly or in groups. The sand and gravel in most kames is at least partially stratified. Kames and kame terraces (Fig. 5-8) are ice-contact deposits showing collapse features where the retaining wall of ice melted.

Ice-channel fillings, another form of glacial stream deposits, comprise eskers (Fig. 5-9) and crevasse fillings. Many eskers have a dendritic pattern reflecting the branching tributaries that were part of the drainage system. Some join or cross one another at right angles and probably represent fillings in reticulate systems of crevasses. Some ice-channel fillings can be traced to kames or kamefields. All these features are surrounded by younger, finer-grained outwash of the later facies of the sequence.

Relict Ice Caps in the Northeastern Highlands (not shown on the map)

As the ice gradually melted back from the northeastern states, ice caps were preserved on the higher mountains, and these relict masses formed active glaciers that moved downslope. One example is on the north slope of the Adirondacks, where rock fragments were transported downward (Denny, 1974, p. 1). The principal highland areas that developed mountain glaciers were in the Catskills and Adirondacks in New York, the Green Mountains in Vermont, the White Mountains in New Hampshire, and the uplands in western and central Maine. In the Adirondacks the ice cap lasted until 12,000 radiocarbon years

Figure 5-7. Glacio-fluvial gravel near Plattsburg, New York. (Photograph by Wayne Newell)

Figure 5-8. Kame terrace, a glacial stream deposit along the edge of a valley. It is composed of cross-bedded sand and gravel. The location is along the Cache la Poudre River, Colorado.

Figure 5-9. Pleistocene glacial drainage pattern in eastern Maine as recorded by the system of eskers (dotted lines). (After Leavit and Perkins, 1935)

ago, by which time the main sheet of ice had retreated into the St. Lawrence Lowland. In the White Mountains and in Maine, the ice caps lasted until 11,500 radiocarbon years ago, by which time the main ice sheet had retreated across the St. Lawrence Lowland (Prest, 1969). Similar ice caps persisted in mountains of the Maritime Provinces in Canada. Apparently, something like 6,000 to 7,000 years elapsed while the Wisconsinan ice sheet retreated from Long Island to the St. Lawrence Lowland.

Glacial Deposits in New Jersey, Pennsylvania, and South and Central New York

West of Long Island, a prominent end moraine can be followed across northern New Jersey, where it crosses the Watchung Ridges, to the Delaware River about 15 miles (24 km) below the Delaware Water Gap. The morainal ridge is poorly developed in eastern Pennsylvania. Stones in the ridge are fresh, like those in the Long Island moraines, and many can be identified as having sources to the north. South of the morainal ridge are deeply weathered till and other clearly pre-Wisconsinan deposits.

The glacial ice in New York moved southward from the lowland bordering Lake Ontario onto the northern edge of the Appalachian Plateau. In doing so it crossed belts of bedrock having different amounts of limestone, and the composition of the till closely reflects the different kinds of bedrock. These deposits apparently contain a high percentage of subglacial till and not much supraglacial till; they differ from the proportions found in New England, probably because of the very different topography crossed by the ice. In northern Pennsylvania, the southernmost Wisconsinan till overlaps deeply weathered pre-Wisconsinan residuum (not shown on the map).

In westernmost New York the rim of the Appalachian Plateau is about 1,200 feet (366 m) higher than Lake Erie and about 5 miles from the lake shore. The late Wisconsinan ice rose out of the lake bed, overrode the rim, and extended about 30 miles (48 km) southeastward across the surface of the plateau, where it built a terminal moraine, locally known as the Kent Glaciation. On its recession a discontinuous end moraine, was deposited on the rim. The lowland bordering the lake is mantled with shore deposits and bottom sediments of Pleistocene lakes that were ancestral to Lake Erie (see Chapter 6).

GLACIAL DEPOSITS IN THE CENTRAL UNITED STATES (MOSTLY TS)

In the central United States the Wisconsinan ice advanced not along a broad front but in a series of lobes that were guided in large part by preexisting topography. There were five major lobes: the Erie, Huron-Saginaw Bay, Lake Michigan, Lake Superior, and Des Moines lobes. Each of these in turn divided into smaller lobes.

The Erie lobe, for example, divided into six lobes. From east to west they are the Grand River, Kilbuck, Sciota, and Miami lobes in Ohio; the East White lobe in Indiana; and the Erie-Maumee lobe in northeastern Ohio and northwestern Indiana. These glacial deposits are intertongued with interstadial deposits. The southern limit of the Wisconsinan drift in western Ohio and eastern Indiana is estimated to be about 17,500 radiocarbon years old. In 3,500 radiocarbon years the ice had retreated to the Lake Erie basin (Prest, 1969).

The Saginaw Bay lobe deposited a series of end moraines (em on the 1/7,500,000 map). In western Indiana and central Illinois they abut the series of moraines deposited by the Lake Michigan lobe. The Lake Michigan lobe had two branches, the Green Bay lobe on the west and a much smaller one at Grand Traverse lowland near the north end of the Michigan peninsula. The Lake Michigan lobe took about 6,000 radiocarbon years to retreat from its maximum limit to a position opposite Green Bay in the north part of the lake basin. The large moraines in Illinois have been assigned to the Wisconsinan.

The various lobes and branches advanced and retreated at different times and rates, and as a consequence the glacial stratigraphy, even of the Wisconsinan, is vastly more complicated than has generally been realized. An example has been described by Wright (1964) in Minnesota, where "the complex interaction of four ice lobes with drifts of contrasting lithology (from east to west the Superior, Rainy, Wadena, and Des Moines lobes). . .allow the identification of five main phases of ice advance."

All these Wisconsinan deposits are young enough to be characterized by well-preserved, original deposition forms. Weathering is slight, and the landforms have been modified only slightly by erosion. End moraines, drumlins, kames, and a few eskers are readily identified in these deposits. Early Wisconsinan deposits are more eroded and more weathered than younger deposits.

Pre-Wisconsinan Deposits in the Central United States

A nearly continuous belt of Illinoian drift more than 50 miles wide is overlapped by and lies south of the Wisconsinan drift in the lowland reaching from Illinois to central Ohio, and a narrow band is exposed on the Appalachian Plateau south of the Kent Till in Pennsylvania and easternmost Ohio. A large number of measurements across the contact between the Wisconsinan and Illinoian drifts in Indiana show an average depth of leaching of 40 to 60 inches in the Wisconsinan drift and an average twice to three times that depth in the Illinoian (Thornbury, 1940).

The Illinoian drift forms a smooth plain of deeply weathered till mantled by loess. The drift covers the entire southern part of Illinois and forms a strikingly different landscape from that of the Wisconsinan drift. It is without end moraines or other constructional landforms typical of younger glacial deposits, and the surface layers are so deeply weathered that hard rocks are decomposed to clay (Chapter 2). Whatever was the course of the Mississippi River during Illinoian time, it now follows the western edge of the Illinoian drift. Buried channel deposits recording abandoned drainages courses (gal on the map) are described in Chapter 7.

West of the Mississippi River, in southern Iowa and northern Missouri, is Kansan drift that is deeply dissected by drainage radiating from the end moraines of the Des Moines lobe. These deposits, too, lack glacial landforms and are more dissected and more deeply weathered than the Illinoian. They are blanketed by loess. The Nebraskan till, known only from subsurface localities, underlies the Kansan.

According to Ray (1966, p. B91), three tills of pre-Wisconsinan age are present south of the Ohio River in Kentucky. The oldest and most weathered (Nebraskan?) caps remnants of a preglacial upland. The youngest (Illinoian?) occurs within the Ohio River valley.

DRIFTLESS AREA

The Driftless Area of southern Wisconsin and northwestern Illinois is certainly not without glacial drift, if by that term we include glacial outwash. Large numbers of glacially striated cobbles have been found there, and some geologists have interpreted them as relics from local till. However, no undisputed deposits of till have been described, and the cobbles could well be residues from outwash. Along the cliffs are boulder fields and masses of rubble that reflect severe periglacial conditions, presumably Wisconsinan. The lowlands and slopes are marked by all sorts of fragile rock monoliths (chimney rocks, Fig. 5-10), delicate stone arches, pedestal rocks, and natural bridges that could not have survived glaciation; and glaciers that extended into that area must have been at least as old as Kansan. Presumably the Driftless Area escaped being glaciated because the ice could move more easily along the valleys now occupied by Lakes Michigan and Superior. The Driftless Area in northern Wisconsin is a thousand feet higher than Lake Superior.

Figure 5-10. Fragile stacks of rock in the Driftless Area of Wisconsin indicate no Late Pleistocene glaciation in that area. Associated with the stacks are pedestal rocks, natural bridges, and arches. (Photograph by John E. Costa)

GLACIAL DEPOSITS OF THE DAKOTA LOBE AND THE GREAT PLAINS

At the Dakota lobe, in the James River Lowland and westward onto the Great Plains, the limits of glaciation and of the moraines is approximately along the Missouri River. Indeed, the course of the river was controlled by the glaciers. Isolated erratics occur west of the river, but for practical purposes the drift ends at the river. The land surface rises westward from the James River lowland, from about 1,300 feet (396 m) altitude in the lowland to almost 4,000 feet (1,219 m) near the foot of the northern Rocky Mountains. Except for small areas of Illinoian drift in southeasternmost South Dakota, the deposits are entirely Wisconsinan in age (Fig. 5-11).

Lemke and others (1965, p. 20) recognized several Wisconsinan ice advances in the region but designated them by numbers rather than trying to correlate them with the deposits farther east, partly because the Middle West nomenclature has become confused and partly because the radiocarbon dates were found to be erratic. There was weathering prior to the deposition of the youngest moraine that loops around the south end of the James River Lowland, and this weathered zone, which is commonly referred to as the Brady Soil, has been used farther west to separate early from late Wisconsinan deposits.

Farther west the Wisconsinan ice moved around the Cypress Hills of southern Alberta and around the isolated laccolithic mountains in Montana. Distinctive trains of boulders from these hills and mountains extend southward from them. Glacial lakes and stream diversions associated with the glacial deposits in North Dakota and Montana are described in Chapters 6 and 7.

GLACIAL DEPOSITS IN THE WESTERN STATES

In the Rocky Mountains of Canada a Wisconsinan ice cap pushed lobes southward into the United States. One of these dammed Clark Fork impounding a lake, Lake Missoula, which overflowed the Spokane River valley and swept across the loess-covered lavas of the Columbia Plateau. The resulting catastrophic floods produced the scablands of eastern Washington (see Chapters 6 and 7).

Two Wisconsinan ice advances in eastern Washington deposited the Spokane and Okanogan tills (Early and Late Wisconsinan, respectively). The ice crossed the canyon of the Columbia River and turned it southward at Moses Coulee and Grand Coulee. The Columbia River, undoubtedly larger than today because of the addition of glacial meltwaters, and further enlarged by floods from Lake Missoula, developed erosion features on so large a scale that geologists long resisted attributing them to stream processes.

Still farther west, two Wisconsinan tills were deposited by lobes of ice that descended the Puget Trough (Crandell, 1965; Easterbrook, 1963), the Early Wisconsinan Admiralty Drift, and the Late Wisconsinan Vashon Drift. These drift sheets were pushed onto the east and west flanks of the northern Cascade Mountains, the upper parts of which were being eroded by alpine glaciers.

South of the limits of the continental ice sheets, snow accumulated on the high mountains sufficiently to form mountain glaciers. Their deposits are shown on the map by mg.

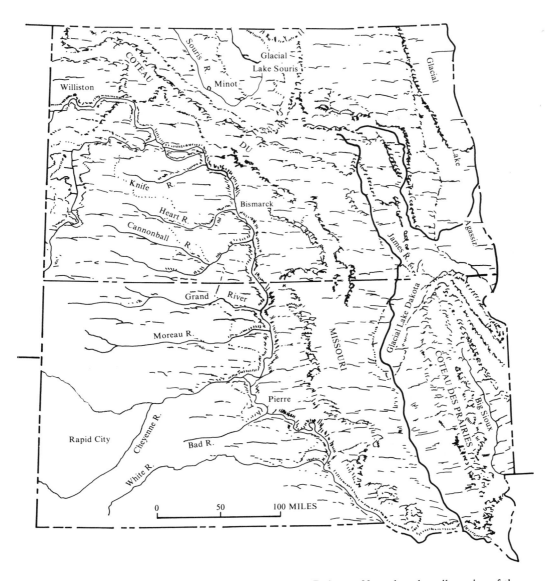

Figure 5-11. Northwest-trending end moraines in the Dakotas. Note that the tributaries of the Missouri River are from the west. The Missouri and its tributaries formerly continued northeastward across the Dakotas to an ancestral river buried under the deposits of Lake Agassiz. Glacial ice diverted that drainage southward. The 1/7,500,000 map shows the buried channels of the former drainage.

Some of the glaciers developed from ice caps on very high plateaus, like the High Plateaus of Utah. Ice lobes from the cap extended off the edges of the plateaus and down the steep valleys descending from the rims. Most of the mountain glaciers, however, headed in semicircular basins known as cirques (Fig. 5-12, see also 6-4) and carved U-shaped valleys below the cirques (Fig. 5-13). Although the front of a glacier may appear stationary, or even be retreating, the ice up-valley continues to move towards the front to replace ice

Figure 5-12.   Cirques in basalt opposite Mt. Marvine in the High Plateaus, central Utah. Altitude is about 11,000 feet (3,353 m).

Figure 5-13.   U-shaped glacial valley along the Roaring Fork, just below Independence Pass, in the Colorado Rockies.

that is melting. Accordingly, debris is continually added to the front of the glacier, partly because of the melting and partly because minor, episodic readvances produce effects like that of a steam-shovel.

Associated with glacial deposits on mountains are residual deposits produced by vigorous frost action, which produces patterned ground (Fig. 2-10) and boulder fields. Because of the small scale of the map, these deposits are included in the pattern mg.

Today, depending on exposure, the lower limit of perennial snowbanks rises southward from approximately 6,500 feet (1,981 m) in the northern mountains to about 12,500 feet (3,810 m) in the southern ones. There was similar change with latitude during the Pleistocene. Near the Canadian border, for example, the mountain glaciers descended to the plains and abutted against the edge of the continental ice sheets. In Colorado the mountain glaciers descended only halfway to the plains. Still farther south the glaciers extended only a short distance below the summits. Figures 5-14 and 5-15 compare terminal moraines in the north with some in Colorado.

Figure 5-14. Lake contained by horseshoe-shaped end moraine at base of the Teton Range, Wyoming.

Figure 5-15. End moraine (E) and right (R) and left (L) lateral moraines at Horseshoe Park, in Rocky Mountain National Park, Colorado. View is down valley.

REFERENCES CITED AND ADDITIONAL BIBLIOGRAPHY

Glacial Deposits, General

Belcher, D. J., and others, 1943, The formation, distribution, and engineering characteristics of soils, Purdue Univ. Res. Ser.,n. 87 and Highway Res. Bull. no. 10, 384p.

Charlewsorth, J. K., 1957, (reprinted) 1966, *The Quaternary Era, with Special Reference to Its Glaciation*, St. Martins Press, New York, 1700p.

Embleton, C., and C. A. M. King, 1968, Glacial and Periglacial Geomorphology, Edward Arnold, London, and St. Martins Press, New York, 608p.

Flint, R. F., 1957, *Glacial and Pleistocene Geology*, Wiley, New York, 353p.

Prest, V. K., 1969, Retreat of Wisconsin and Recent ice in North America, *Geol. Survey of Canada Map 1257A*, scale 1/5,000,000.

U.S. Depprtment of Agriculture, 1938, *Soils and Men*, Yearbook.

Wright, H. E., Jr., 1964, The classification of the Wisconsin glacial stage, *Jour. Geology* **72:**628-637.

Wright, H. E., Jr., and D. G. Frey, eds., 1963, *The Quaternary of the United States*, Princeton University Press, Princeton, N.J., 932p.

Glacial Deposits, Eastern States

Alden, W. C., 1925, The physical features of central Massachusetts, *U.S. Geol. Survey Bull. 760*, pl. XIII

Antevs, E., 1922, The recession of the last ice sheet in New England, *Am. Geogr. Soc. Res. Ser. No. 11*, 188p.

Borns, H. W., Jr., 1963, Preliminary report on the age and distribution of the late Pleistocene ice in north central Maine, *Am. Jour. Sci.* **261:**738-749.

Chute, N. E., 1959, Glacial geology of the Mystic Lakes—Fresh Pond area, Massachusetts, *U.S. Geol. Survey Bull. 1061-F*, pp. 87-216.

Coates, D. R., 1966, Glaciated Appalachian Plateau: till shadows on hills, *Science* **152:**1617-1619.

Denny, C. S., 1956, Wisconsin drifts in the Elmira region, N.Y. and their possible equivalents in New England, *Am. Jour. Sci.* **254:**82-95.

Denny, C. S., 1974, Pleistocene geology of the northeast Adirondack region, New York, *U.S. Geol. Survey Prof. Paper 786*, 50p.

Denny, C. S., and W. H. Lyford, 1963, Surficial geology and soils of the Elmira-Williamsport region, New York and Pennsylvania, *U.S. Geol. Survey Prof. Paper 379*, 60p.

Donner, J. J., 1964, Pleistocene geology of eastern Long Island, New York, *Am. Jour. Sci.* **262:**356-376.

Flint, R. F., 1930, The glacial geology of Connecticut, *Connecticut State Geol. and Nat. Hist. Survey Bull. 47*, 294p.

Flint, R. F., 1953, Probably Wisconsin substages and late Wisconsin events in northeastern United States and southeastern Canada, *Geol. Soc. America Bull.* **64:**897-919.

Fuller, M. L., 1914, The geology of Long Island, *U.S. Geol. Survey Prof. Paper 82*, 231p.

Goodlett, J. C., 1954, Vegetation adjacent to the border of the Wisconsin drift in Potter Co., Pa., *Harvard Forest Bull. 25*, 93p.

Goodlett, J. C., 1956, Vegetation and surficial geology, in Denny, C. S., Surficial Geology and Geomorphology of Potter Co., Pa., *U.S. Geol. Survey Prof. Paper 288*, pp. 56-66.

Hartshorn, J. H., 1958, Plowtill in southeastern Massachusetts, *Geol. Soc. America Bull.* **69:**477-482.

Holmes, G. W., 1966, Stephen Reed, M. D., and the "celebrated" Richmond boulder train of Berkshire County, Massachusetts, U.S.A., *Jour. Geology* **6**:431–437.

Jahns, R. H., 1951, Surficial geology, Mt. Toby Quadrangle, Mass., U.S. Geol. Survey GQ-9.

Jahns, R. H., 1953, Surficial geology, Ayer Quadrangle, Mass., U.S. Geol. Survey GQ-21.

Jahns, R. H., 1966, Surficial geologic map of the Greenfield Quadrangle, Franklin County, Mass., U.S. Geol. Survey GQ-474.

Johnson, F. (ed.), The Boylston Street Fishweir (correct?) II, Peabody Foundation Archeology Papers, vol. 4 (1), 133p.

Kaye, C. A., 1964, Outline of Pleistocene geology of Martha's Vineyard, Mass., *U.S. Geol. Survey Prof. Paper 501-C*, pp. C134–139.

Leavitt, H. W., and E. H. Perkins, 1935, Glacial geology of Maine, *Maine Tech. Exp. Sta. Bull. 30*, vol. 2, 232p.

Lockwood, W. N., and H. Meisler, 1960, Illinoian outwash in southeastern Pennsylvania, *U.S. Geol. Survey Bull. 1121-B*, 9p.

MacClintock, P., 1954, Leaching of Wisconsin glacial gravels in eastern North America, *Geol. Soc. America Bull.* **65**:369–384.

MacClintock, P., and J. Terasmae, 1960, Glacial history of Covey Hill, New York—Quebec, *Jour. Geology* **68**:232–241.

Rich, J. L., 1935, Glacial geology of the Catskills, New York *State Mus. Bull. 299*, 180p.

Schafer, J. F., and J. H. Hartshorn, 1965, The Quaternary of New England, in *The Quaternary of the United States*, H. E. Wright, Jr., and D. G. Frye, eds., Princeton University Press, Princeton, N.J., pp. 113–128.

Upson, J. E., and C. W. Spencer, 1964, Bedrock valleys of the New England coast as related to fluctuations of sea level, *U.S. Geol. Survey Prof. Paper 454-M*, 44p.

Glacial Deposits, Central States

Alden, W. C., 1953, Physiography and glacial geology of eastern Montana and adjacent areas, *U.S. Geol. Survey Prof. Paper 231*, 200p.

Bergstrom, R. E., ed., 1968, The Quaternary of Illinois, *University of Illinois College of Agriculture Spec. Publ. 14,* 179p.

Bretz, J. H., 1955, Geology of the Chicago region Pt. II. The Pleistocene, *Illinois Geol. Survey Bull 65*, 132p.

Condra, G. E., and others, 1950, Correlation of the Pleistocene deposits of Nebraska, *Nebraska Geol. Survey Bull. 15-A*, 74p.

Flint, R. F., 1955, Pleistocene geology of eastern South Dakota, *U.S. Geol. Survey Prof. Paper 262*, 173p.

Frye, J. C., and H. B. Wilman, 1960, Classification of the Wisconsin Stage in the Lake Michigan glacial lobe, *Ill. Geol. Survey Circ. 285*, 16p.

Goldthwaite, R. P., and others, 1961, Glacial map of Ohio, U.S. Geol. Survey Misc. Geol. Inv. Map I-316.

Goldthwaite, R. P., and others, 1965, Pleistocene deposits of the Erie lobe, in *The Quaternary of the United States*, H. E. Wright, Jr., and D. G. Frye, eds., Princeton University Press, Princeton, N.J., pp. 85–97.

Horberg, C. L., and R. C. Anderson, 1956, Bedrock topography and Pleistocene glacial lobes in central United States, *Jour. Geol.* **64**:101–116.

Hough, J. L., 1958, *Geology of the Great Lakes*, University of Illinois Press, Urbana, 313p.

Lemke, R. W., and others, 1965, Quaternary geology of northern Great Plains, in *The Quaternary of the United States*, H. E. Wrig‚ht, Jr., and D. G. Frey, eds., Princeton University Press, Princeton, N.J., pp. 15–27.

Leverett, F., 1902, Glacial formations and drainage features of the Erie and Ohio Basins, *U.S. Geol. Survey Mon. 41*, 802p.

Leverett, F., 1929, Moraines and shorelines of the Lake Superior region, *U.S. Geol. Survey Prof. paper 154-A*, 72p.

Leverett, F., 1932, Quaternary geology of Minnesota and parts of adjacent states, *U.S. Geol. Survey Prof. Paper 161*, 149p.

Martin, L. M., 1916, The physical geology of Wisconsin, *Wis. Geol. Nat. Hist. Survey Bull. 36*, 547p.

Ray, L. L., 1966, Pre-Wisconsin glacial deposits in northern Kentucky, *U.S. Geol. Survey Prof. Paper 550-B*, pp. B91-94.

Ruhe, R. V., and others, 1957, Late Pleistocene radiocarbon chronology in Iowa, *Am. Jour. Sci.* **265:**671-689.

Thornbury, W. D., 1940, Weathered zones and glacial chronology in southern Indiana, *Jour. Geology* **48:**449-475.

White, G. W., 1951, Illinoian and Wisconsin drift of the southern part of the Grand River lobe in eastern Ohio, *Geol. Soc. America Bull.* **62:**967-977.

White, G. W., 1960, Classification of Wisconsin glacial deposits in northeast Ohio, *U.S. Geol. Survey Bull. 1121-A*, 12p.

White, G. W., 1961, Classification of glacial deposits in the Killbuck Lobe, northeast-central Ohio, *U.S. Geol. Survey Prof. Paper 424-C*, pp. 71-73.

Glacial Deposits in Western States

Alden, W. C., 1953, Physiography and glacial geology of western Montana and adjacent areas, *U.S. Geol. Survey Prof. Paper 231*, 200p.

Atwood, W. W., and K. F. Mather, 1932, Quaternary geology of the San Juan Mountains, Colorado, *U.S. Geol. Survey Prof. Paper 166*, 176p.

Atwood, W. W., 1909, Glaciation of the Unita and Wasatch Mountains, *U.S. Geol. Survey Prof. Paper 61*, 96p.

Birkeland, P. W., 1964, Pleistocene glaciation of the northern Sierra Nevada north of Lake Tahoe, California, *Jour. Geology* **72:**810-825.

Birman, J. H., 1964, Glacial geology across the crest of the Sierra Nevada, *Geol. Soc. America Spec. Paper 75*, 80p.

Bretz, J. H., and others, 1956, Channeled scabland of Washington: new data and interpretations, *Geol. Soc. America Bull.* **67:**957-1049.

Crandell, D. R., 1965, The glacial history of western Washington and Oregon, in *The Quaternary of the United States*, H. E. Wright, Jr., and D. G. Frey, eds., Princeton Univ. Press, Princeton, N.J., pp. 341-353.

Dickson, R. G., 1945, Landslide origin of the type Cerro till, southwestern Colorado, *U.S. Geol. Survey Prof. Paper 525-C*, pp. C147-C151.

Easterbrook, D. J., 1963, Late Pleistocene glacial events and relative sea-level changes in the northern Puget Lowland, Washington, *Geol. Soc. America Bull.* **74:**1465-1484.

Fryxell, F. M., 1930, Glacial features of Jackson Hole, Wyoming, *Augustana Library Publ. 13*, 128p.

Horberg, C. L., and R. A. Robie, 1955, Post-glacial volcanic ash in the Rocky Mountain piedmont, Montana and Alberta, *Geol. Soc. America Bull.* **66:**949-955.

Matthes, F. E., 1930, Geologic history of Yosemite Valley, *U.S. Geol. Survey Prof. Paper 160*, 137p.

Moss, J. E., 1951, Late glacial advancesd in the southern Wind River Mountains, Wyoming, *Am. Jour. Sci.* **249:**865-883.

Putnam, W. C., 1949, Quaternary geology of the June Lake district, California, *Geol. Soc. America Bull.* **60:**1281-1302.

Ray, L. L., 1954, Glacial chronology of the southern Rocky Mountains, *Geol. Soc. America Bull.* **51:**1851-1917.

Richmond, G. M., 1962, Quaternary stratigraphy of the LaSal Mountains, Utah, *U.S. Geol. Survey Prof. Paper 324*, 135p.

Shroder, J. F., and R. E. Sewall, 1985, Mass movement in the LaSal Mountains, Utah, in Contributions to Quaternary geology of the Colorado Plateau, G. E. Christenson, ed., *Utah Geol. and Mineral Survey Spec. Studies 64*, pp. 49-85.

Smith, J. F., Jr., and others, 1963, Geology of the Capitol Reef area, Wayne and Gardfield Counties, Utah, *U. S. Geol. Survey Prof. Paper 363*, 102p.

Spieker, E. M., and M. P. Billings, 1940, Glaciation in the Wasatch Plateau, Utah, *Geol. Soc. America Bull. 51*, 1173-1198.

Weis, P. L., and G. M. Richmond, 1965, Maximum extent of late Pleistocene Cordilleran glaciation in northeastern Washington and northern Idaho, *U.S. Geol. Survey Prof. paper 525C*, pp. C128-C132.

LAKE DEPOSITS (1 ON THE MAP)

Lakes were greater in size and number during Pleistocene times than they are today. They were of many kinds, depending on whether they formed:

1. In front of the continental ice sheets as a result of rivers being dammed by ice (Fig. 6-1, Table 6-1);
2. In basins and river valleys overdeepened as a result of scour by ice (in part the Great Lakes; also, the Finger Lakes of New York);
3. In closed depressions on ground moraine (Fig. 6-2);
4. In structural basins, like the more than 100 closed basins in the Basin and Range Province;
5. In sinkholes where formations are readily soluble, like limestone sinks in Florida, southern Indiana, (Fig. 3-16) and western Kentucky and mostly dry gypsiferous and limestone sinks in southeastern New Mexico's karst land and on the Kaibab Plateau, Arizona;
6. Along rivers at cut-off meanders (oxbow lakes) or depressions in deltas;
7. Back of landslides;
8. Back of lava flows, in volcanic craters; or
9. In depressions that result in part from wind erosion, as on the Great Plains.

In addition, in today's world there are artificial lakes back of dams.

Table 6-1. Altitudes and maximum depths of the Great Lakes

Lake	Altitude of water surface	Maximum depth	
		Feet	Meters
Superior	602	1333	406
Michigan	580	923	281
Huron	580	750	228.6
Erie	572	210	64
Ontario	246	778	237

GREAT LAKES (TABLE 6-1) AND OTHER GLACIAL LAKES IN THE CENTRAL UNITED STATES

Lakes and ponds and the sediments laid down in them are second only to glacial drift as characteristc of glaciated regions. The northeastern and north-central parts of the United States abound with lakes and ponds, including the Great Lakes, and these are only the southern part of a vast expanse of lake country that extends northward for hundreds of miles into the Canadian Arctic. Besides the Great Lakes, other large lakes in a belt that extends north across Canada include Lake Winnipeg, Lake Athabasca, Great Slave Lake, and Great Bear Lake.

The lakes have pollution problems. Lake Erie is the worst, perhaps because its volume is comparatively small. Dumping waste into the lakes has been based on the dictum that the "solution to pollution is dilution," but the dumping has become excessive. Moreover, the bottom sediments are accumulating toxic substances.

The Great Lakes were formed when Pleistocene glaciers retreated from that area, and several of the lakes have rounded ends reflecting their glacial history. Figure 6-1 illustrates six stages in the development of the lakes and the river channels by which they overflowed. The dried parts of the lake beds form some of our finest agricultural land, like that of Lake Maumee at the western end of Lake Erie and of glacial Lake Saginaw at the head of Saginaw Bay in Michigan. Between these plains of lake sediments is hummocky morainal topography with deposits of countless other small lakes (Fig. 6-2).

Similar ice-marginal lakes must have formed in front of pre-Wisconsinan glaciers, but little record of them remains. One in southeastern Ohio represented by the Minford Silts of Nebraskan or Kansan age seems to have formed when the drainage was northward across the site of the Ohio River (Stout and Schaaf, 1931). Another lake along the Ohio River formed at Paducah (Olive, 1966).

Niagara Falls came into existence when retreat of glacial ice allowed the level of lake water in the Ontario Basin to fall below the limestone rim of the Niagara Escarpment. Thereafter, Lake Erie drained into Lake Ontario via the Niagara River and Niagara Falls. Because of undercutting of shale that underlies the limestone rim at the falls, the falls have receded about 4 feet (1.2 m) per year, a distance of 7 miles (11.3 km) upstream. The rate of retreat, though, has not been constant; part of the time Lake Huron and other lakes to the west drained to the Mississippi River. There were stages (Fig. 6-1) when Lake Huron and the other lakes to the west drained to the St. Lawrence River valley by way of the Ottawa River. At such times the Niagara drained only Lake Erie.

Earth movements caused by isostatic adjustment to loading and unloading of the ice are recorded today by tilting of the shorelines (Fig. 6-3) of the ancestral Great Lakes. Such movements, in fact, continue today. The north shores of Lake Superior and Lake Ontario are rising northward at the rate of 6 to 12 inches (152.4 to 304.8 mm) per 100 miles (161 km) per century because of unloading of the Canadian ice sheets.

Low divides between tributaries of the Great Lakes and the Ohio and Mississippi rivers facilitated French exploration of the Middle West. Easy portages could be made, and one could travel by canoe from Quebec to New Orleans. Later, canals were dug along some of the portage routes that had been followed by early travelers.

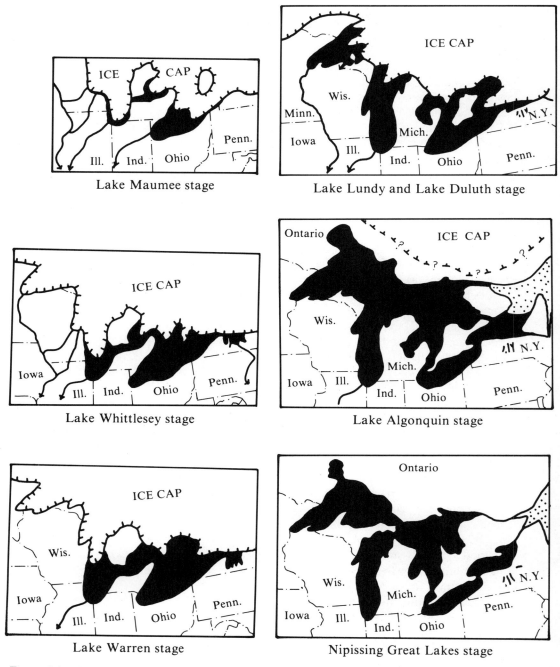

Figure 6-1. Maps showing stages in development of the Great Lakes. (Based on Leverett and Taylor, 1915, and Leverett, 1928)

Figure 6-2. Lakes, ponds, and marshes in depressions in hummocky morainal topography south of Jackson, Michigan. Some of the marshes shown here have been dredged or filled, but such ground remains poorly drained. Contour interval is 10 feet (3.1 m). (From U.S. Geological Survey Jackson 15-minute quadrangle, 1939)

GLACIAL LAKE DEPOSITS, EASTERN UNITED STATES

Among the larger glacial lakes in New England was glacial Lake Hitchcock, which formed in the Connecticut River Valley (1 on the map) when the valley was dammed by a moraine near Middletown, Connecticut. The moraine was deposited during a minor readvance of the retreating ice sheet. This lake extended as a single body of ponded water northward into southern Canada

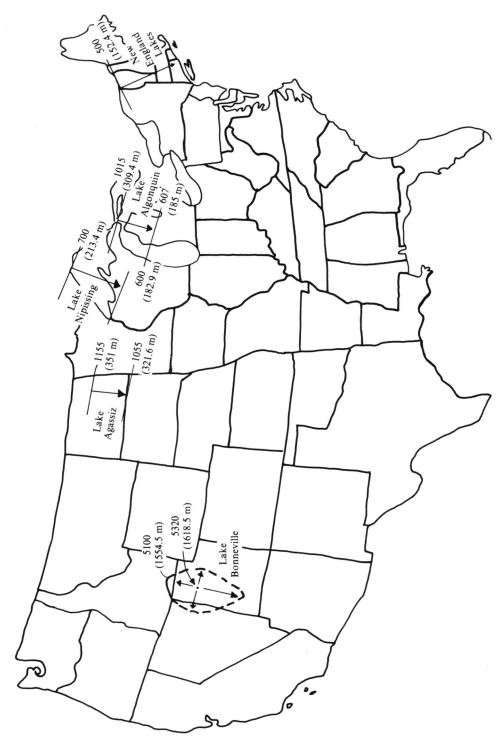

Figure 6-3. Tilting of Late Pleistocene shorelines due to isostatic rebound. Numbers show present elevations (feet above sea level) along shorelines of Wisconsinan-age lakes. In the northeastern and north-central states, the northern shorelines have risen (and are still rising) higher than the southern shorelines, following unloading of the continental crust by the melting of glacial ice. In Utah, unloading by loss of water from Pleistocene Lake Bonneville has caused the central part of its basin to dome upward (Fig. 6-6).

(Stewart and MacClintock, 1969, pp. 102-103). Varve counts by Antevs (1922) indicate that retreat of ice from Middletown, Connecticut, to St. Johnsbury, Vermont, required 4,100 years. Radiocarbon dates indicate that the lake existed for 2,300 years (Schafer and Hartshorn, 1965, p. 122). Stewart and MacClintock (1969, p. 103) question the varve dating because the ice retreat was by stagnation, but the radiocarbon dates are also suspect because the samples were obtained from open sites where biochemical contamination almost certainly occurred.

In most places the sediments of Lake Hitchcock rest on bedrock rather than on older till. Due to isostatic rise of the land, the beach lines rise northward at about 4.15 feet per mile, from 435 feet at the south to 1,100 feet at the Canadian border. As in the Great Lakes region, the shorelines tilt southward.

Among the numerous smaller glacial lakes in New England are several that developed in the Nashua River valley, northwest of Worcester, Massachusetts. Another lake, glacial Lake Passaic, developed back of Watchung Ridges between Newark and Morristown, New Jersey, when Wisconsinan ice was at its maximum extent and built the end moraine (em) that extends west from Long Island.

The history of the latest Wisconsinan and Early Holocene lake and glacial deposits in the Adirondacks and bordering areas of Vermont and Quebec involved (see MacClintock and Stewart, 1965; Stewart and MacClintock, 1969; Denny, 1974):

1. Development of a succession of ice-dammed lakes in the Finger Lakes region as the late Wisconsinan ice retreated into Quebec, but the St. Lawrence Lowland continued to be dammed by ice;
2. Readvance of ice in Vermont and onto the Adirondacks;
3. Retreat of this ice, producing ice-dammed lakes that lengthened northward as the streams draining the Adirondacks were ponded;
4. Formation of ancestral Lake Champlain (Lake Vermont) when ice dammed the Champlain Valley. Lake Iroquois, which developed in the front of the ice west of the Adirondacks, overflowed eastward across a bedrock ridge at Covey Hill, near the north base of the Adirondacks, cut a gorge 300 feet (91.4 m) deep in the bedrock, and discharged into the Hudson River;
5. Further withdrawal of the ice front to lower Lake Iroquois, which merged with Lake Vermont about 12,200 radiocarbon years ago;
6. When the ice front had retreated to a position a short distance north of Quebec City, there was a marine submergence (Champlain Sea) that lasted from about 12,000 to about 10,500 radiocarbon years ago;
7. Development of the freshwater Lake Champlain when the land gradually rose (isostatically) above sea level, about 10,000 radiocarbon years ago.

Near the north edge of the Appalachian Plateau are a dozen or so nearly parallel, north trending lakes, the Finger Lakes of New York, mentioned above. They are attributed mostly to ice scour of preglacial valleys in Devonian shale, but near the north rim of the plateau they are eroded in limestone. The lakes are located at altitudes from slightly less than 400 (121.9 m) to more than 1,000 feet (304.8 m). The bottoms of the two largest, Seneca Lake and Lake Cayuga, are below sea level. Near the south end of Lake Cayuga, wells indicate that the valley fill is more than 600 feet (182.9 m) deep. Whether the preglacial valleys drained northward to Lake Ontario and the St. Lawrence River or southward to the Susquehanna River is not known. The lakes are contained in their southern ends by a moraine (Valley Heads Moraine) generally regarded as late Wisconsinan in age and correlated by Denny (1956) with the moraines on Long Island.

GLACIAL LAKE DEPOSITS, WESTERN UNITED STATES

Glacial lake deposits in the western United States include those of Lake Agassiz (1) in eastern North Dakota and western Minnesota. This was the most extensive of all the glacial lakes in North America. Other lake deposits formed in Montana when the continental ice sheets dammed the tributary valleys of the Missouri River.

The deposits of glacial Lake Agassiz form a plain covering more than 10,000 square miles in the Red River Valley along the North Dakota-Minnesota border. The plain is about 800 feet (243.8 m) in altitude near the border with Canada; the eastern side of the valley rises to about 1,200 feet (365.8 m) onto the morainic topography formed by the glacial lobes that advanced into western Minnesota, and the western side rises more than that to the edge of the Great Plains. At its maximum extent the lake overflowed southward via the valley of Traverse Lake to the valley of the Minnesota River. This outlet is presently at an altitude of just under 1,000 feet (304.8 m).

Part of the time Lake Agassiz discharged southward, but lower outlets opened eastward at times as the Minnesota lobes of glacial ice retreated, and these would be closed again by minor readvances. The first stages of Lake Agassiz ended when withdrawal of the ice opened outlets to the east, but readvance of the ice in the latest Wisconsinan times again raised the lake level to the level of the outlet at the south. Retreat of the ice during earliest Holocene times reopened drainage outlets to the east. Lake history ended roughly 8,200 radiocarbon years ago, when the retreat of the ice reached the Nelson River, north of Lake Winnipeg, and the drainage could then escape to Hudson Bay.

The Lake Agassiz deposits, which form one of the largest areas of level land in the world, are in three belts. In the center of the valley are clays and clay loams, difficult to plow and very poorly drained, yet highly productive, especially of hay. Around the center is a belt of silt loams or very fine sandy loams that are very productive of grains. Around this is a third belt composed of sandy loam and sand along the old beach lines.

Deposits of two other lakes in North and South Dakota occur in the James River Basin (Lake Dakota) and along the Souris River (Lake Souris when the drainage was dammed by retreating ice). These deposits are mostly silt loam.

Farther west, in Montana, southward advance of the ice sheets pushed the Missouri to a new course south of its original course, dammed the streams tributary from the south, and developed lakes in the tributary valleys. These lake deposits, except at the west, are in narrow valleys incised into the Missouri Plateau, and the deposits are not extensive.

One of the most important lakes in terms of shaping the landscape was Lake Missoula in western Montana. Its importance stems from the fact that it repeatedly overflowed its ice dam to produce catastrophic floods with volumes as great as 50 cubic miles of water that formed the scablands on the Columbia basalts in eastern Washington. Lake Missoula covered about 3,300 square miles (8,547 sq. km) and at its ice dam had a maximum depth of 2,000 feet (609.6 m). Shorelines are distinct but closely spaced, as if the lake level fluctuated greatly. Quantities of silt were washed into the lake, in places 100 feet (30.5 m) thick. North of Lake Missoula another considerable lake formed in the Kootenai River valley.

Farther west is yet another glacial lake, Lake Chelan, in a valley deepened by ice scour. Both on the east side of the Cascades and in the Puget Trough, southward advance of the ice lobes dammed the valleys draining into the lowlands and developed lakes in most of the valleys. These were drained when the ice melted. At Seattle are two glacial lakes, Lake Washington about 15 miles (24 km) long paralleling Puget Trough, and the smaller Sammamish Lake about 3 miles (4.8 km) farther east. These lake basins, like Puget Trough itself, were deepened by ice scour.

The lake deposits shown on the map in the western part of the Snake River plain, in east-central Washington and north-central Oregon, are pre-Wisconsinan deposits laid down when the drainage was ponded by folding of the lavas. Lake deposits in southeastern Oregon result in part from lava dams and in part from volcanic craters.

The environmental effects of the very different kinds of ground (lake beds, till, and glacial outwash) in the Seattle area include major differences in the stability of the ground, differences in depth to the water table, differences in drainage conditions, suitability for excavation, and availability of sand and gravel. Lake beds form the least stable slopes (McKee, 1972, p. 299).

Glacial Lake Deposits on the Western Mountains

Glacial lake deposits on the western mountains are numerous, widely distributed, but individually small and thin. Many cirques have small lakes or ponds, and the sediments in them are Holocene. A few glaciated valleys have kame terraces deposited when water was ponded along the edge of the valley ice. Ponded sediments grading into stream-deposited outwash are common back of many terminal moraines in the glaciated valleys (Fig. 5-15). Some terminal moraines that were built onto the plains at the foot of the Teton Mountains contain lakes (Fig. 5-14). Still other glacial lakes are located in undrained depressions in ground moraine on some of the plateaus that were ice-capped during the Wisconsinan glaciation, such as those on Grand Mesa, Colorado (Yeend, 1969, p. 25), and on the Aquarius Plateau in the High Plateaus of Utah (Flint and Denny, 1958).

An unusual combination of a lake in a cirque ponded by postglacial faulting (Fig. 6-4) is at Snow Lake in the Wasatch Plateau (Spieker and Billings, 1940, Pl. 4 and p. 1192).

Lake Deposits in the Basin and Range Province

Another type of lake that formed during Pleistocene times includes those of the Great Basin and other parts of the Basin and Range Province. These first

Figure 6-4.   Lake in cirque closed by fault (bluff at upper left) across mouth of the cirque. Snow Lake in the Wasatch Plateau, about 10 miles (16.1 km) east of Manti, Utah.

became well known through the classic reports a century ago by G. K. Gilbert on Lake Bonneville and by I. C. Russell on Lake Lahontan and Mono Lake. Lakes were numerous in those desert basins and formed at the time of the glaciations in the western mountains. Glacial outwash from the mountains intertongues with the lake deposits.

Lake Bonneville was a vast lake that covered 20,000 square miles (51,800 sq. km) in northwestern Utah and reached a maximum depth of 1,000 feet (304.8 m) (Fig. 6-5). It was a body of water comparable in size to Lake Michigan. The ecology of that desert region was very different, and the lake period ended only about 10,000 years ago.

The Lake Bonneville deposits include deltas covering several square miles at the mouths of rivers, and these became the sites of most of Utah's largest cities. Between the deltas are conspicuous shorelines, embankments, bars, and spits (Fig. 6-6). Offshore from these are deepwater facies of silt and clay. The highest shoreline, the Bonneville Beach, in the words of G. K. Gilbert, "Shouts for recognition," for it marks a topographic unconformity separating the V-shaped valleys and gullies on the mountain slope above from the predominately horizontal beach features below.

When the lake reached its maximum height, it overflowed into the Snake River (Fig. 6-7), cut a gorge in landslide and alluvial deposits at the pass where it overflowed, and then maintained the level of the pass in bedrock while the Provo Beach and formation was deposited about 325 feet (99.1 m) below the Bonneville Beach. The large deltas deposited during this flooding still stand.

Thereafter the lake gradually desiccated and fell 650 feet (198.1 m) to the level of the Great Salt Lake, the near-salt-saturated remnant of the once-vast freshwater Pleistocene lake. A beach line about midway between the Provo Beach and the Great Salt Lake, known as the Stansbury Beach, is generally regarded as the terminal stage of Lake Bonneville. Pleistocene vertebrates are

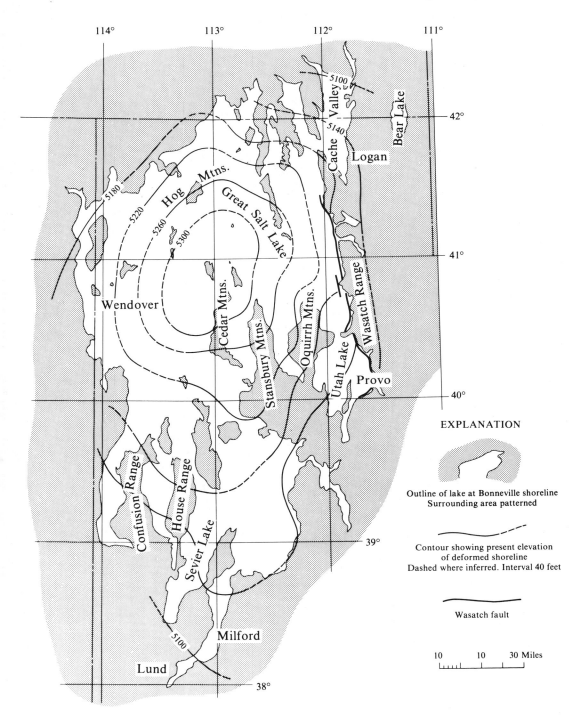

Figure 6-5. Map showing extent of Lake Bonneville, the ancestral Great Salt Lake, and the isostatic doming of its shorelines by the drying of the lake. Faults indicated are those that disrupt the high shoreline. (From Crittenden, 1963, but see also Gilbert, 1890)

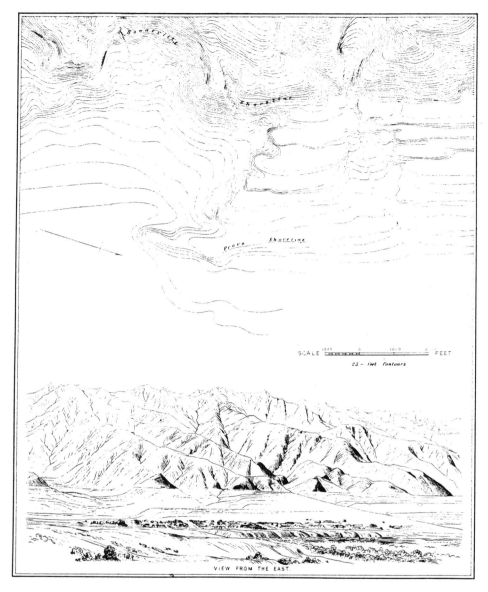

Figure 6-6. Example of shore features of Pleistocene Lake Bonneville. Location at Wellsville, Utah. A striking topographic unconformity separates the horizontal lines of the lake deposits from the steep valleys carved by streams on the mountainside. As Gilbert stated, "It shouts for recognition." (From Gilbert, 1890).

found in old deposits; they are not found in deposits younger than the Stansbury Beach.

The glacial deposits that intertongue with the Lake Bonneville deposits represent the early and late Wisconsinan glaciations. Drill records in the Lake Bonneville basin reveal older Pleistocene lake beds interbedded with fan gravels and other stream deposits that are important aquifers in the now-desert basins. Beneath the Pleistocene fill in the valleys is Tertiary fill, locally more than 10,000 feet (3,048 m) thick, containing considerable admixed volcanic debris.

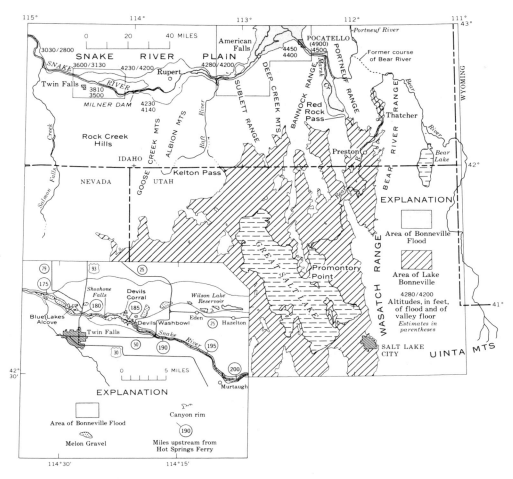

Figure 6-7. Map of the northern part of Lake Bonneville, showing Red Rock Pass where the lake overflowed and the path of the Bonneville Flood along the Snake River Plain. (From Malde, 1968)

Lake Bonneville was extensive enough and deep enough to become a major localized load on that part of the continental crust. It bent downward under the load and then rebounded as the load was removed. Figure 6-6 shows the extent of the lake and how the shorelines have warped upward isostatically as a result of removal of the load of water.

Except for the Bonneville Flood, the history of Lake Bonneville is largely duplicated at other large, Late-Pleistocene lakes in the Great Basin, notably Lake Lahontan and Mono Lake. Moreover, the several lakes were probably contemporaneous because they all are correlative with the Late Pleistocene (Wisconsinan) glaciations.

Another feature of the Lake Bonneville deposits, and of those at the other lakes and playas in the Great Basin, are evaporites. As the lakes dried, their dissolved salts precipitated in orderly zones reflecting differences in the solubilities of the different salts. In general, there is a peripheral zone of carbonate salts, an intermediate zone dominated by sulfate salts, and an inner zone of chlorides (Fig. 6-8). In the Great Salt Lake desert, as a result of isostatic rise of the basin,

Figure 6-8. The history of a saltpan in the Great Basin, as illustrated by a salty residue in an evaporating dish. *A*: Evaporation of a brine from the dish deposits various salts in an orderly sequence determined by their solubilities: **c**, carbonates; **s**, sulfates; and **h**, chlorides. *B*: As evaporation progresses, the saltpan may become tilted and the salt zones become overcrowded to one side. *C*: New additions of fresh water (dark rivulets) by rains and floods rework the salts into irregular patterns. (Diagrams by John R. Stacy, in Hunt et al., 1966)

the rings of salts are crowded westward. In Death Valley, because of tilting that accompanied faulting of that valley floor, the salts are crowded eastward.

The lake basin at Death Valley and others now marked by playas in the Mojave Desert were once connected by through-drainage tributary to the Colorado River. The Amargosa River, Death Valley, Owens River, and the Mojave River originally were an integrated system draining southeast, perhaps via Bristol, Cadiz, and Danby lakes (Fig. 6-9) to join the Colorado River (Hunt, 1969, p. 123). The evidence for this is a fish story; species of fish living in now-isolated springs in those deserts are related to species in the Colorado River (Hubbs and Miller, 1948, pp. 88-90). Playas are ringed by alluvial fans. Ground-water, checked by the fine-grained playa sediments, is ponded to form a border zone of springs at the edge of the playa.

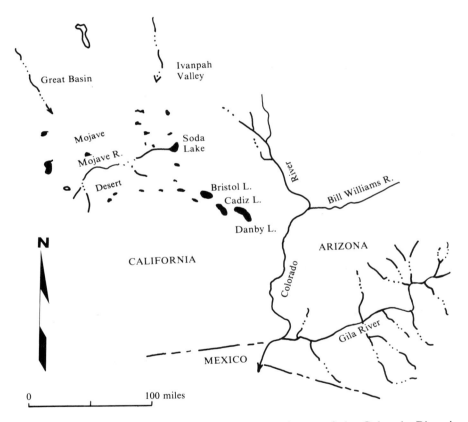

Figure 6-9. Contrast in drainage systems east and west of the Colorado River in southern California and southern Arizona. In southern California the drainage has been disrupted by faulting and folding. The Mojave River formerly discharged to the Colorado River via Bristol, Cadiz, and Danby Dry lakes, and it received drainage from the Great Basin, including the Owens River, Death Valley, and the Amargosa River (shown diagrammatically). The courses of these former tributaries also have been disrupted. East of the Colorado River, in Arizona, long tributaries from the east maintained their courses. (After Hunt, 1969)

OTHER KINDS OF LAKES

Other kinds of lakes and lake beds generally are small. One common kind is found in karst terrain, where formations are readily soluble—for example, the numerous lakes in Florida, where limestone roofs over cavernous limestone have collapsed, and karst ground with sinks in the limestone of the Kaibab Plateau on the north rim of Grand Canyon, and in the limestone plateaus in western Kentucky and southern Indiana (Fig. 3-16), some of which provide underground courses for rivers. Karst ground is extensive, too, in the limestone and saliferous formations of southeastern New Mexico.

Still other kinds of lakes, also mostly small, develop in cut-off meanders at river beds. Mark Twain, commenting on how the Mississippi River had shortened its course by cutting off meanders, predicted that New Orleans and St. Louis will someday be joined.

Earthquake Lake along the Madison River gorge through the Madison Range formed back of the Madison Canyon landslide (Fig. 3-10). Reelfoot Lake, in Tennessee at the head of the Mississippi Embayment, is another earthquake lake, formed in 1811-1812 when the land there subsided 3 to 9 feet (0.91 to 2.7 m) during the New Madrid earthquake.

Crater Lake, Oregon, exemplifies lakes in volcanic craters or calderas. The lakes on Newberry Volcano, also in Oregon, are another example. Related lakes, but much smaller and more ephemeral, develop at maar volcanoes (steam-explosion craters) such as Hole in the Ground and Big Hole south of Newberry Volcano, and Kilbourne and Hunt's Hole west of Las Cruces, New Mexico. Zuni Salt Lake in New Mexico may be another, although it has also been explained as the result of collapse over dissolved salts.

Malheur and Harney lakes in eastern Oregon are examples of lakes produced by lava dams. These lakes formerly discharged to the Snake River by way of the Malheur River, but a lava flow dammed the outlet and they now flow only in wet times. Bear Lake in northwestern Utah formerly drained to the Portneuf River, but lava flows ponded the river and turned it southward to Great Salt Lake.

Older lake deposits were laid down by the Snake River where it enters Hell Canyon at the west end of the Snake River plains. Presumably the river was ponded by episodic uplift of the barrier through which Hell Canyon is eroded.

REFERENCES CITED AND ADDITIONAL BIBLIOGRAPHY

Alden, W. C., 1925, The physical features of central Massachusetts, *U.S. Geol. Survey Bull. 760-B*, pp. 13-105.

Antevs, E., 1922, The recession of the last ice sheet in New England, *Am. Geog. Soc. Research Ser. No. 11*, 120p.

Antevs, E., 1948, The Great Basin, with emphasis on glacial and postglacial times; Three climatic changes and pre-white man, *Utah Univ. Bull.* **38**(20):168-191.

Bissell, H. J., 1963, Lake Bonneville: Geology of southern Utah Valley, Utah, *U.S. Geol. Survey Prof. Paper 257-B*, pp. 101-130.

Craig, R. C., 1984, *Climates and lakes of the Death Valley drainage system during the last glacial maximum*, Dept. Geology, Kent State University, 157p.

Craig, R. C., 1984, *Analysis of ice-age flooding from Lake Missoula*, Dept. Geology, Kent State University, 147p.

Crippen, J. R., and B. R. Pavelka, 1970, The Lake Tahoe Basin, California—Nevada, *U.S. Geol. Survey Water-Supply Paper 1972*, 56p.

Crittenden, M. D., Jr., 1963, New data on the isostatic deformation of Lake Bonneville, *U.S. Geol. Survey Prof. Paper 454-E*, 31p.

Darton, N. H., and others, 1908, Passaic, N.J.—N.Y., *U.S. Geol. Survey Folio 157.*

Denny, C. S., 1956, Wisconsin drifts in the Elmira region, N.Y. and their possible equivalents in New England, *Am. Jour. Sci.* **254:**82-95.

Denny, C. S., 1974, Pleistocene geology of the northeast Adirondack region, New York, *U.S. Geol. Survey Prof. Paper 786*, 50p.

Elson, J. A., 1957, Lake Agassiz and the Mankato-Valders problem, *Science* **126:**999-1002.

Feth, J. H., and others, 1966, Lake Bonneville: Geology and hydrology of the Weber River delta district, *U.S. Geol. Survey Prof. Paper 518*, 76p.

Flint, R. F., and C. S. Denny, 1958, Quaternary geology of Boulder Mountain, Aquarius Plateau, Utah, *U.S. Geol. Survey Bull. 1071-D*, pp. 103-164.

Gilbert, G. K., 1890, Lake Bonneville, *U.S. Geol. Survey Monograph I*, 438p.

Gwynn, J. W., 1980, Great Salt Lake: a scientific, historical, and economic overview, *Utah Geol. and Mining Survey*, 400p.

Hough, J. L., 1958, *Geology of the Great Lakes*, University of Illinois Press, Urbana, Ill., 313p.

Hubbs, C. L., and R. R. Miller, 1948, The zoological evidence, *Utah Univ. Bull.* **38**(20):18-166.

Hunt, C. B., and others, 1966, Hydrologic basin, Death Valley, California, *U.S. Geol. Survey Prof. Paper 494-B*, 138p.

Hunt, C. B., and others, 1953, Lake Bonneville: geology of northern Utah, *U.S. Geol. Survey Prof. Paper 257-A*, 99p.

Hunt, C. B., 1969, Geologic history of the Colorado River, in The Colorado River Region and John Wesley Powell (Powell Memorial Volume), *U.S. Geological Survey Prof. Paper 669*, pp. 59-130.

Johnston, E. A., 1946, Glacial Lake Aggasiz, with special reference to the mode of deformation of the beaches, *Geol. Survey Canada Bull. 7*, 29p.

Kotell, C., 1963, Glacial lakes near Concord, Mass., *U.S. Geol. Survey Prof. Paper 475-CV*, pp. 142-144.

Lane, R. K., and D. A. Livingstone, 1974, Lakes and lake systems, in *Encyclopedia Britannica*, vol. 10, pp. 600-616.

Leverett, F., 1928, Moraines and shorelines of the Lake Superior Basin, *U.S. Geol. Survey Prof. Paper 154-A*, 72p.

Leverett, F., and F. B. Taylor, 1915, The Pleistocene of Indiana and Michigan and the history of the Great Lakes, *U.S. Geol. Survey Monograph 52*, 641p.

MacClintock, P., and D. P. Stewart, 1965, Pleistocene geology of the St. Lawrence Lowland, *New York State Mus. and Sci. Service Bull. 394*, 152p.

Malde, H. E., 1968, The catastrophic late Pleistocene Bonneville flood in the Snake River Plain, Idaho, *U.S. Geol. Survey Prof. Paper 596*, 52p.

McKee, B., 1972, *Cascadia, The Geologic Evolution of the Pacific Northwest*, McGraw-Hill, New York, 394p.

Morrison, R. B., 1964, Lake Lahontan; Geology of southern Carson Desert, Nev., *U.S. Geol. Survey prof. Paper 401*, 156p.

Olive, W. W., 1966, Lake Paducah, of late Pleistocene age, in western Kentucky and southern Illinois, in Geological Survey Research, 1966, *U.S. Geol. Survey Prof. Paper 550-D*, pp. D87-D88.

Putnam, W. C., 1949, Quarternary geology of the June Lake district, California, *Geol. Soc. America Bull.* **60:**1281-1302.

Putnam, W. C., 1938, The Mono Craters, Calif., *Geog. Rev.* **29**:68-82.

Russell, I. C., 1885, Geological history of Lake Lahontan, a Quaternary lake of north-western Nevada, *U.S. Geol. Survey Monograph 11*, 288p.

Schwartz, G. M., and G. A. Theil, 1954, Minnesota's Rocks and Waters, University of Minnesota Press, Minneapolis, Minn. 366p.

Schaffer, J. P., and J. H. Hartshorn, 1965, The Quaternary of New England, in *The Quaternary of the United States*, Princeton Univ. Press, Princeton, N.J., pp. 113-128.

Scott, W. E., and others, 1983, Reinterpretation of the exposed record of the last two cycles of Lake Bonneville, *Quaternary Research* **20**(3):281-285.

Smith, W. O., and others, 1960, Comprehensive survey of sedimentation in Lake Mead, 1948-1949, *U.S. Geol. Survey Prof. Paper 295*, 254p.

Spieker, E. M., and M. P. Billings, 1940, Glaciation in Wasatch Plateau, Utah, *Geol. Soc. America Bull.* **51**:1173-1197.

Stewart, D. P., and P. MacClintock, 1969, The surficial geology and Pleistocene history of Vermont, *Vermont Geol. Survey Bull. 31*, 251p.

Stout, W. E., and D. Schaaf, 1931, Minford silts of southern Ohio, *Geol. Soc. America Bull.* **42**:663-672.

Williams, J. S., 1962, Lake Bonneville: Geology of southern Cache Valley Utah, *U.S. Geol. Survey Prof. paper 257-C*, pp. 131-152.

Yeend, W. E., 1969, Quaternary geology of the Grand and Battlement Mesa area, Colorado, *U.S. Geol. Survey Prof. Paper 617*, 50p.

STREAM DEPOSITS AND HISTORIES
OF SOME DRAINAGE SYSTEMS

Deposits laid down by streams have many forms. Some are alluvium (al) deposited along floodplains, including some buried under glacial deposits (dotted lines on the map) that identify abandoned stream courses. Or the alluvium may be buried under lava flows, like the lavas in southeastern Idaho that dammed Bear River and turned it into the Great Salt Lake. Other streams build fan-shaped deposits of gravel (fg), sand (fs), and silt (not shown separately on the map) at the foot of mountains. Fan deposits are especially characteristic of the structural basins in the Southwest, where they form long gravelly surfaces (bajadas) sloping from the base of the mountains to the floors of the basins, which may be floodplains or dry lakes (playas).

The surfaces of alluvium and fan deposits are constructional, and the deposits, especially the fans, generally are thick. A third category of stream deposits texturally resembles the constructional fills but forms a comparatively thin mantle over previously eroded surfaces. The eroded surfaces may be terraces cut on rock along the edges of valleys (rock-cut terraces), or they may be pediments that slope from the foot of hills or mountains. The surface forms of these deposits on erosion surfaces are indistinguishable from constructional fills; rock-cut terraces mantled by alluvium are superficially like alluvial fills, and pediments mantled by gravel superficially look like fan gravels, but the gravels on pediments are thin. One has to look inside the terrace or fan in order to distinguish it.

Drainage systems and their stream deposits develop patterns that are characteristic of the structure of the watershed. Drainage systems are likely to be parallel (consequent drainage) where there is a long, even slope, as, for example, along the Atlantic Coastal Plain or the western slope of the Sierra Nevada. On isolated mountains, like volcanoes and mountains attributed to intrusions on the Montana Plains, Colorado Plateau, and west Texas, the drainage patterns are radial. In closed basins, like that of the Great Salt Lake and other basins of the Southwest, the drainage is centripetal.

Many streams have tributaries that join at acute angles and so form a dendritic pattern. These may branch like an elm, as do the headwaters of the

Missouri River in Montana, or like an oak, as does the Mississippi. Other dendritic patterns may have tributaries on only one side, like a flag tree at a wind blasted timberline; the Missouri River in the Dakotas and the Colorado River below Hoover Dam are examples. In the Appalachians, between the Blue Ridge and Allegheny plateaus, streams are parallel between linear ridges but make right-angle turns through watergaps to join an adjacent valley, thus forming a trellis pattern.

Hydrologists distinguish between: a) bedload, the material rolled along the bottom of a channel; b) suspended load, the sediment carried in suspension by the turbulent water; and c) dissolved load, the material carried in solution. If the flow becomes slackened, whether by natural or artificial causes, part of the bottom load becomes deposited and part of the suspended load becomes bottom load. Conversely, when velocity is increased part of the bottom load may become suspended.

Some streams flow down the dip of the rock formations, (consequent streams) like the parallel streams on Tertiary formations of the Great Plains; others flow against the dip, like the Colorado River where it crosses the Colorado Plateau. John Wesley Powell, who made the first scientific exploration and study of the Colorado River (1869-72), distinguished between longitudinal valleys that parallel the structure and transverse valleys that cut across the structure; he described three varieties of each (Fig. 7-1).

Powell wondered how transverse river valleys could get across structural barriers, and he offered two explanations of how the Green River, as an example, could get across the Uinta Mountains. One hypothesis he termed antecedence (Powell, 1875, p. 163). According to this interpretation, a stream maintains its course, while a mountain barrier is uplifted slowly across the river's course; down-cutting keeps pace with uplift. Such a stream is older than the uplifted barrier.

The second hypothesis, which he referred to as superposition (Powell, 1875, p. 166), assumes a transverse barrier buried under younger deposits. A stream developed on those young deposits accidentally encounters the temporarily quiescent barrier but is able to maintain its course across the barrier by cutting downward into it. Such a stream is younger than the exhumed barrier. He concluded that the Green River must be antecedent, but this cannot be because the river is on Miocene formations before it turns into its gorge in Precambrian rocks, which were uplifted before the Miocene, at the Gates of Lodore. But the river cannot be superimposed either, because that would require excessive fill north of the Uinta Mountains.

A combination of Powell's two hypotheses, referred to as anteposition, meets the difficulties that generally are not accounted for by either of the hypotheses alone. Anteposition assumes that a stream course became established by superposition during a stage when uplift was still not high, and the stream then deepened its canyon by antecedence as later stages of uplift resumed at the barrier. The Colorado River, which developed out of Early Tertiary lakes on the Colorado Plateau, was superimposed across the pre-Eocene (Laramide) structures, but the river and its Grand Canyon are antecedent across the mid-Tertiary and later displacements that included regional uplift of the plateau.

transverse valleys

longitudinal valleys

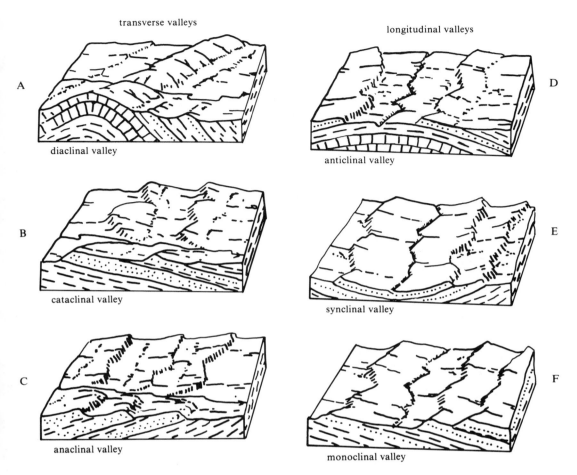

A

diaclinal valley

B

cataclinal valley

C

anaclinal valley

D

anticlinal valley

E

synclinal valley

F

monoclinal valley

Figure 7-1. Types of valleys as related to geologic structure, as distinguished by John Wesley Powell. Transverse valleys are at right angles to the strike of the rock structures and cut across them. There are three varieties: some cross anticlinal uplifts (*A*); others are transverse to the flanks of broad folds and may flow in the direction of the dip (*B*); or against the dip (*C*). Longitudinal valleys parallel the strike of the structure, and these may be along the crest anticlines (*D*), along the trough of a syncline (*E*), or along the flanks of the broad folds (*F*).

G. K. Gilbert (1877, p. 144) stated that drainage may become superimposed in several ways: (1) from marine deposits that bury old structures and then become elevated and dissected (e.g., the Coast Ranges of Oregon); (2) from alluvial or other terrestrial deposits that bury old structures and then become dissected (e.g., the gorge of the Rio Grande through the Pliocene or Pleistocene lava flow at San Acacia, New Mexico); and (3) from erosion surfaces truncating older structures (e.g., the eastward drainage across the Great Plains).

The foregoing relates drainage systems to structural forms and structural history. Also important are the effects of climate, and especially climatic change. In general, wet periods during the Pleistocene favored alluviation; dry episodes favored erosion and even dune development. Too, there are differences in streams attributable to seasonal distribution of precipitation. Along the Pacific Coast precipitation occurs mostly during the winter and causes floods and landslides; in the central United States precipitation occurs mostly during the summer, when growing plants help retard erosion. Also important are the extremes—extremes of drought that dry up streams and extremes of snowfall or snow melt that cause flash floods.

Finally may be mentioned the changes in stream sedimentation and erosion caused by artificial works; excess runoff from urban areas is one example. Another is the erosion of channel beds that commonly develops immediately downstream from newly built dams, the deposition of sediments in the upper ends of the lakes back of dams, and the flood disasters caused when dams fail.

ALLUVIUM AND ALLUVIAL TERRACES, WESTERN UNITED STATES (AL)

Quaternary alluvial deposits, geology's ubiquitous Qal, are widespread and provide a clear record of the change from Pleistocene to Holocene vertebrate faunas. Figure 7-2 illustrates some different kinds of relationships found in alluvial deposits. Generally the oldest alluvium contains remains of Pleistocene animals, such as mammoth, camel, horse, and long-horned bison. At a few

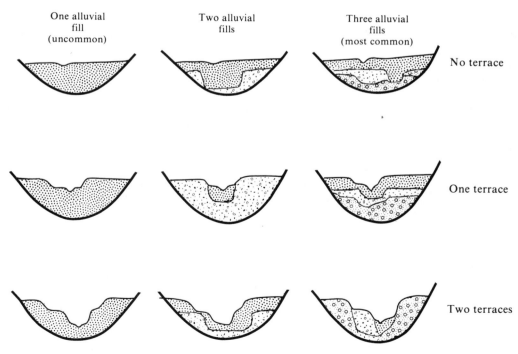

Figure 7-2. Examples of valley cross sections showing some common relationships between alluvial fills and their associated alluvial terraces. (After Leopold and Miller, 1954)

localities these are associated with the remains of early man. Younger alluvial deposits contain a Holocene fauna, but these can commonly be distinguished by their archaeological stratigraphy. In the western United States, for example, Holocene alluvium older than about A.D. 1 is characterized by preceramic artifacts; younger alluvium contains fragments of pottery and artifacts associated with the ceramic occupations.

The Late Pleistocene, mid-Holocene, and late Holocene alluvial deposits are separated by unconformities marked by arroyos partly or wholly filled with the next younger alluvium and/or sand dunes. Clearly, the periods between the alluviations were comparatively dry times.

In the Rocky Mountain region the alluvial deposits have pretty generally been cut by arroyos (Figs. 7-3, 7-4). That overgrazing was a major cause seems indicated by the fact that the erosion began at different times in different valleys and in each valley began shortly after settlement. However, similar erosion attributable to climatic change has occurred in the past, so we cannot be sure whether overgrazing was the cause or merely hastened the inevitable. Moreover, the dendrochronology record for the Southwest generally indicates that the latter half of the nineteenth century was a drier than average period.

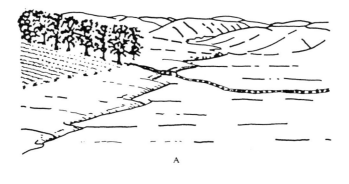

A

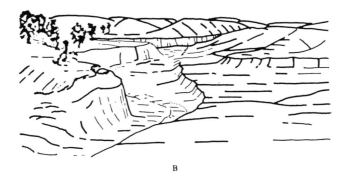

B

Figure 7-3. A valley on the Colorado Plateau before arroyo cutting (*A*) and after (*B*). Location is Pleasant Creek in the western part of the Henry Mountains area. (After Hunt et al., 1953)

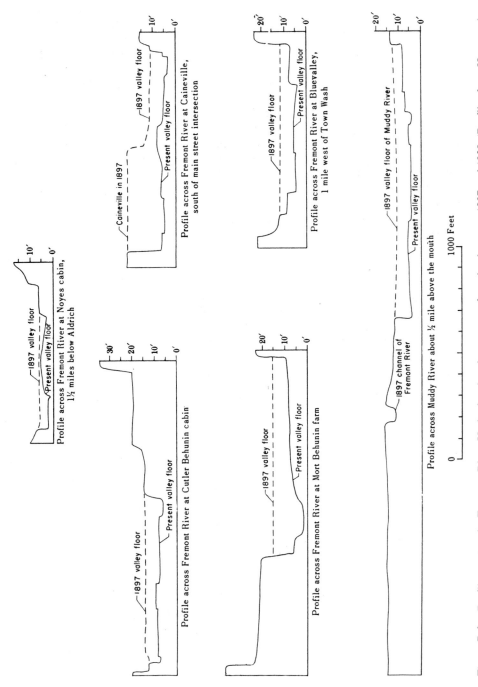

Figure 7-4. Profiles across the Fremont River, Utah, showing the amount of erosion between 1897 and 1938. (Profiles from Hunt et al., 1953)

ALLUVIAL FANS AND RELATED DEPOSITS ON PEDIMENTS (FG, FS, OSG)

Fan deposits, particularly those in the Basin and Range Province, can be considered in four groups. The largest, in both area and volume, are composed of gravel (fg) grading downslope to sand and silt (fs). These fans are typically several miles long and 500 to 1,000 feet (152.4 to 304.8 m) high (Fig. 7-5). Geophysical surveys indicate that many fans like those illustrated are more than a mile thick. Probably only about a quarter of such deposits are Quaternary; the remainder is Tertiary, and much of it is volcanic in origin.

A second group of gravel fans may be equally long and high but have their surfaces interrupted by hills and ridges of older rocks protruding through the gravel. The rocks being eroded in the mountains may be similar to those contributing to the big fans, but the fan deposits are thinner. Still a third group of fans, illustrated by those along the east side of Death Valley (Fig. 7-6*A*), are small because the valley floors have been down-faulted. Such fans have been sinking and are partly buried by overlap of the playa sediments. At the New-

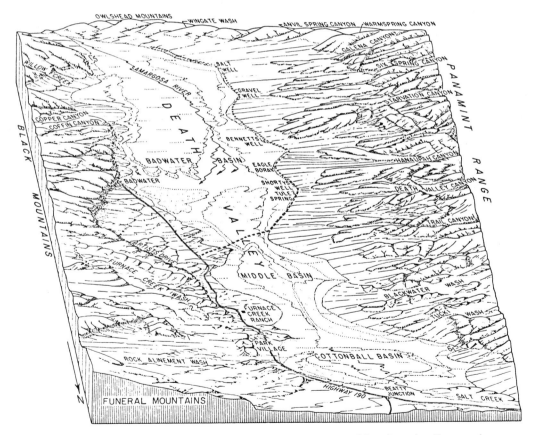

Figure 7-5. Block diagram of Death Valley, showing gravel fan deposits. Fans at the foot of the Panamint Range are 5-6 miles (8-9.7 km) long and 1,000 feet (304.8 m) high. Those along the east side of the valley are smaller because the valley floor has been tilted eastward and the lower parts of those fans are buried under playa sediments on the valley floor. See also Figure 7-7. (From Hunt and Mabey, 1966)

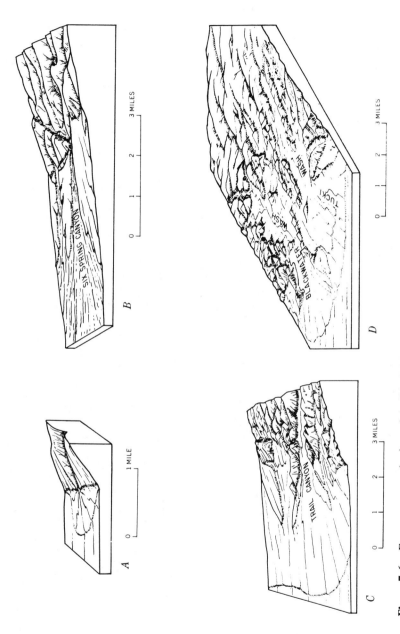

Figure 7-6. Fan patterns at the foot of the Black Mountains and Panamint Range. *A*: Fan-based west front of the Black Mountains (at Coffin Canyon). View north. *B*: Fan-based east foot of the Panamint Range at Six Spring Canyon. View south. *C*: Fan-frayed east foot of the Panamint Range at Trail Canyon. View south. *D*: Fan-wrapped east foot of the Panamint Range at Blackwater and Tucki washes. View southwest. The differences in fan patterns are orderly and reflect different rates of deformation along the edges of the mountains. Downfaulting of the valley is most rapid and most recent along the front of the Black Mountains (*A*). The Panamint Range has been tilted north, causing progressively greater overlap of valley fill onto the mountain from *B* to *D*. (From Hunt and Mabey, 1966)

foundland Mountains, near Great Salt Lake, the fans are so completely buried that the sediments of Pleistocene Lake Bonneville overlap the mountainsides.

A fourth group of fans contain a high percentage of fine-grained sediments, either because they are remote from the source rocks, like the fans crossed by Interstate Route 80 west of Deming, New Mexico, or because the source rocks are fine-grained.

The beautifully developed gravel fans in Death Valley also illustrate very well that fan patterns vary depending on their structural history (Fig. 7-6). The asymmetry of the fans on the east and west sides of the valley can be attributed to eastward tilting of the valley floor. Somewhat similar asymmetry, however, may be caused by fine-grained deposits being deposited on one side of a valley while coarse gravels are deposited on the other side.

On many Southwest desert fans the main washes, which are transporting coarse material from the mountains, have steeper gradients than tributaries. The main washes, with their coarse debris, rise in areas of relatively high rainfall but discharge into areas of low rainfall where the ground is permeable and where water is lost by seepage and evaporation. The courses of these streams are being aggraded. Their tributaries, though, rising no higher than the foot of the mountains, are cutting down, especially those eroding in fine-grained rocks; such tributaries may have flatter gradients than the main streams (Figs. 7-7, 7-8, 7-9).

Figure 7-7. Gravel-covered pediments look like fans but have very different origins and different internal structures. This view, along with Figure 7-8, of pediments at the foot of the Henry Mountains, Utah, provides examples of the various stages in the process of erosion of pediments and subsequent covering of the pediments with gravel from the mountains. In this view, Cedar Creek (main arroyo at left) is still perched on the high gravel bench, but streams in the badlands left (north) of the bench have cut far below that stream, and Cedar Creek will be diverted to those streams. When that happens, the gravels being transported by Cedar Creek will be deposited in the shale valleys and will spread across the pediments along them.

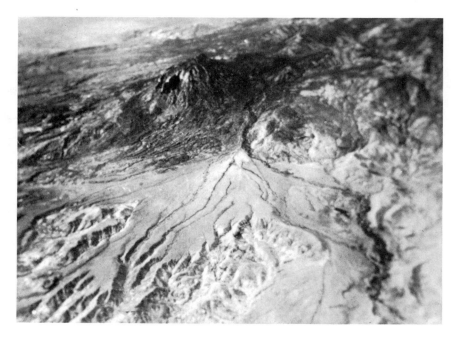

Figure 7-8. Another view (with Fig. 7-7) of the pediments at the foot of the Henry Mountains, Utah, showing a more advanced stage of pedimentation. Bull Creek deposited the gravel capping the high bench at the foot of the mountain, but has been diverted west (right) to the lower courses of tributaries in the shale formations along the west side of the high benches. Compare Figure 7-9.

Figure 7-9. Streams draining from desert mountains transport gravels, and when they discharge onto surrounding deserts they lose their water by seepage and evaporation. The gravels therefore are deposited, and the stream course develops a steeper gradient than its tributaries, which rise in the desert and erode their valleys in shale or other easily eroded rocks. This view is across a desert wash tributary to Bull Creek and just off the lower edge of the view in Figure 7-8. The desert wash, foreground, has been eroded lower than the channel of Bull Creek, which is marked by the row of trees. Clearly Bull Creek will be captured by the desert wash and when turned into the wash will deposit its load of gravel there.

Orderly changes in drainage patterns accompany deposition of gravel fans. A bare rock pediment may have a dendritic drainage pattern that becomes obliterated when a gravel-laden stream is diverted onto the pediment. The aggrading stream has a braided pattern as it is turned first one way then another by old channels becoming clogged. Finally the surface is built sufficiently steeply so new additions of gravel can be moved across it, and the drainage becomes parallel and consequent.

The origin of pediments has been attributed to rill wash and alternatively to lateral planation by the main streams. A part of the difference in opinion can be resolved by more precise definitions that distinguish between pediments representing individual rock fans as distinct from those representing confluent rock fans. On a single rock fan, both processes, and perhaps others like sheet flooding, may operate, but a series of confluent rock fans clearly cannot be attributed to lateral planation by a single master stream. Also, in considering pediment origins it is necessary to determine to what extent the gravels capping the pediment are contemporaneous with it and involved in the process of cutting, and to what extent they are younger and not at all related to the pediment.

Old erosion surfaces at high levels in mountain areas have commonly been interpreted as remnants of former open valley stages representing pauses in the mountain uplifts (Fig. 7-10). A topographic unconformity separates the maturely eroded upper surface from the youthful surrounding topography that slopes steeply from it.

Topographic unconformities also occur where pediments have been eroded on granitic or other crystalline rocks at the foot of mountains formed of the same kind of rocks (Fig. 7-11). Such erosion surfaces are extensive in southern Arizona and the Mojave Desert. At such places the alluvial gravels on the pediment thicken downslope, and the eroded surface of the pediment slopes basinward comparatively steeply beneath increasingly thick basin fill.

In some parts of the Basin and Range Province, like the central part of the Great Basin, the extent of the basins and their fills is only a little greater than the extent of the mountains supplying the fills. These basins are characterized by alluvial fans and almost no pediments. In southern Arizona and the Mojave Desert, on the other hand, the basins are very much more extensive than the mountains, many of which are small and low. These basins have extensive pediments as well as alluvial fans. The differences do not seem to be due to differences in recency of the faulting because basins east and west of the Colorado River have equally well developed pediments.

Alluvial fans are extensive also along the west base of the Sierra Nevada. Those fans have radial profiles that are not smooth curves but consist of three or four straight-line segments, attributed to intermittent uplift of the Sierra (Bull, 1964a, 1964b). Some of these fans are underlain by clay-rich deposits and have subsided sufficiently to damage canals, roads, pipelines, transmission lines, and buildings. The subsidence does not appear to be the result of withdrawal of groundwater but may result from compaction of the clays or, more likely, a process known as piping. Piping involves erosion of the clays by groundwater flowing along the cracks in the clay.

Fans and pediments are extensive along the east base of the Rocky Mountains too. The processes involved appear to be the same as those that have operated where exposures are better along the foot of the Book Cliffs in Utah (Rich, 1935) and around the foot of the Henry Mountains (Figs. 7-7, 7-8). The

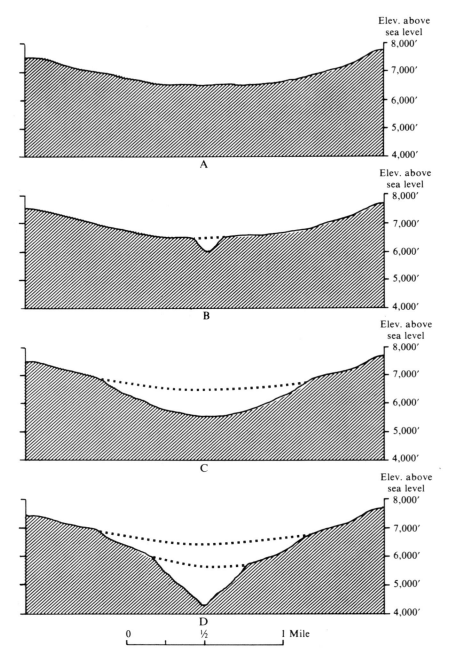

Figure 7-10. Cross sections illustrating successive stages in the cutting of Merced Canyon in the Sierra Nevada. The sections are drawn to scale and show the general proportions of that part of the canyon that ultimately became the Yosemite Valley. The first, or broad valley, stage (*A*) dates back presumably to the Late Miocene when the Sierra Nevada was still relatively low. The inner gorge (*B*) cut by the Merced after the first strong tilting of the Sierra Nevada (*C*) is the second, or mountain valley, stage dating back presumably to the Pliocene epoch. The third, or canyon, stage (*D*) was produced early in the Quaternary period by rapid incisions of the present inner gorge in consequence of the last great Sierra uplift. (After Matthes, 1930)

Figure 7-11. Many pediments of the Mojave Desert and southern Arizona are eroded in granitic or other crystalline rocks that differ little from the rocks in the mountains. *Top* is a view across pediments eroded on granitic rock in the Organ Pipe Cactus National Monument, Arizona. *Bottom* is a close-up view in one of the washes showing about 5 feet (1.5 m) of bedrock under a shallow (1-3 foot) (0.3-0.91 m) cover of subrounded gravel.

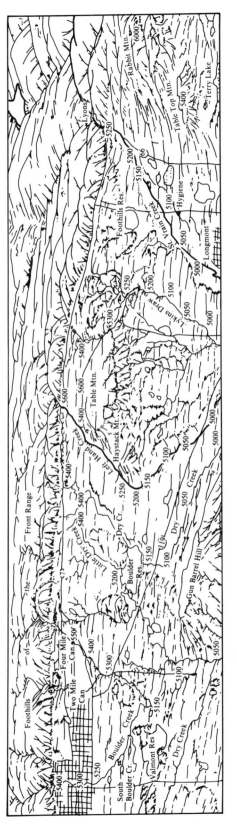

Figure 7-12. Diagram of pediment surfaces along the foot of the Front Range between Boulder and Lyons, Colorado. Altitudes at about 50-foot (15.2 m) intervals are indicated along the principal stream courses. The main streams begin in a high-rainfall area in the mountains and then discharge to an area of less rainfall on the plains. As a consequence, their flow diminishes away from the mountains because of seepage and, to a lesser degree, evaporation. They are therefore depositing rather than eroding. The pediments were probably not produced by lateral planation by those main streams; they were probably eroded by tributaries that are cutting into the shale formations on the plains and that are not overloaded with gravel from the mountains. This process is more pronounced on the Colorado Plateau (cf. Figs. 7-7, 7-8, 7-9), but even here along the foot of the Front Range, some tributaries have lower gradients than the main streams. See, for example, Dry and Little Dry creeks and the small tributariues to Lykins Draw.

incipient capture of Rock Creek at Red Lodge, Montana, practically duplicates the relations shown in Figure 7-8. The same can be seen, although less clearly, in the arrangement of the surfaces and fans along the foot of the Front Range in Colorado (Fig. 7-12).

Gravel-capped pediments like those shown in Figures 7-7 and 7-8 are easily identified, but at many places pediments cannot be distinguished easily from alluvial flats around them. Figure 7-13 is an example along the Canadian River near the New Mexico-Texas boundary. That the differences are important ecologically is indicated by the fact that the two kinds of ground generally have different kinds of vegetation, with the more drought-resistant species growing on the pediment. The differences can be important in some land use problems, such as suitability for agriculture, or for excavation and drainage.

The fan deposits and pediments that have been described are of Quaternary age. Older fan deposits of Pliocene and even older ages form uplands on the Great Plains. The present eastward drainage from the Rocky Mountains began when the mountains began to be uplifted at the end of the Cretaceous period. Although the mountains began to be formed at the end of the Cretaceous, their bases remained low until much later, during the Tertiary, when there was regional uplift as well as continued deformation at the mountains. As regional uplift progressed, fans were built eastward onto the plains, and towards the end of the Tertiary the drainage became incised into those fan deposits (osg in eastern Colorado and New Mexico). The Quaternary fans and pediments at the present foot of the Rockies are a thousand or more feet below the projected positions of those old fans. The streams crossing the plains and the alluvium along their floodplains probably did not extend all the way to the Mississippi River system until Late Tertiary times. The probabilities are that runoff from the mountains discharged only partway down the fans and that water discharging at the toes of the fans was generated by floods originating at or below the middle of the fans, as happens where the extent of fans is very much greater than the extent of storms that produce the floods.

The development of the Pecos River and deposition of alluvium along it (al in southeastern New Mexico) were caused by streams cutting downward into

Figure 7-13. Pediment sloping north from cliffs along the south side of the Canadian River Valley in northwestern New Mexico. Clayey or loamy ground in the Triassic shale formations of the area is of two kinds that are difficult to distinguish because they grade into each other. One kind consists of residual clay or loam on broad low mounds; the other kind consists of thin alluvial loam washed onto pediments eroded on the shale. The smoothness of the foreground here suggests that this pediment is an alluvial surface.

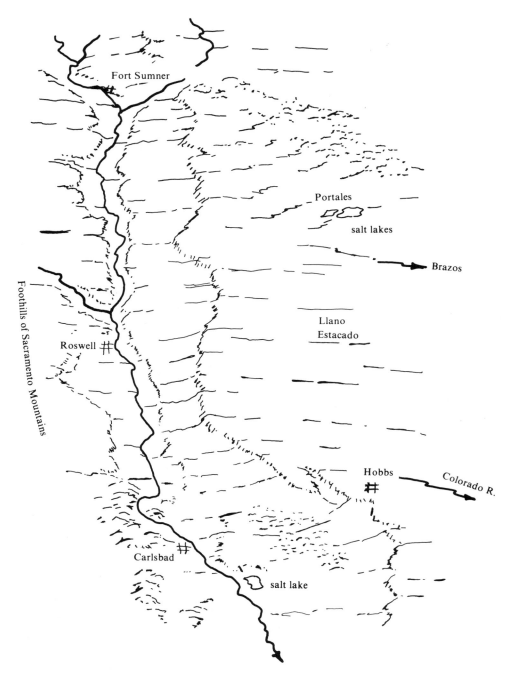

Figure 7-14. The Pecos River rises in the Rocky Mountains and now drains southward across eastern New Mexico to join the Rio Grande in Texas. Originally, it turned eastward near Fort Sumner into the sag at Portales, and it continued eastward, probably during the Early or Middle Pleistocene. Later it was captured and turned southward by the river eroding headward between the Llano Estacado and the Sacramento mountains. Also captured were the headwaters of the Brazos River and the headwaters of the Colorado (of the east) River.

the Pliocene fans (osg). Heads of streams that formerly drained eastward onto the plains were captured and diverted southward by the Pecos River (Fig. 7-14), probably about Kansan times. There were drainage changes on the plains too. The Salina River in Kansas formerly turned southward to join the Smoky Hill River, and the joined streams continued southeastward to join the Arkansas. By Kansan times the streams had become incised 150 feet (46 m) below the surface of the Pliocene fan, and they have subsequently eroded their valleys 200 feet (61 m) lower.

DRAINAGE CHANGES RELATED TO GLACIATION IN THE NORTH-CENTRAL UNITED STATES

The glaciations in the north-central United States caused major changes in the drainage, some of which have already been mentioned, like the course of the Missouri River across North and South Dakota (Fig. 5-11). In those states the tributaries of the Missouri River all flow from the west, but those eastward-flowing streams formerly continued northeastward across what is now the trench of the Missouri River and turned northward into Canada. On the surficial deposits map the old, abandoned courses are marked by valley fills (gal) buried under the glacial deposits.

Other drainage changes shown on the map, marked by deposits designated "gal," are in southeastern South Dakota and along the Mississippi River in southeastern Iowa and western Illinois. Still another major drainage change, partly shown on the 1/7,500,000 map, is the ancient course of the Ohio River, known as Teays, (gal in Ohio, Indiana, and eastern Illinois). An example along the Platte River is shown in Fig. 7-15. All these buried stream channel deposits are important sources of groundwater.

MISSISSIPPI RIVER TERRACES

Four alluvial terraces have been described along the lower part of the Mississippi River valley, but there is major disagreement about how they correlate with the glacio-fluvial deposits in the north. Those who have worked northward from the delta (Fisk and MacFarlan, 1955) correlated the alluvial section with the several glacial and interglacial deposits as follows:

Upper Mississippi Valley	Lower Mississippi Valley
Holocene and Late Wisconsinan	alluvium of the floodplain
Late Wisconsinan	valley cutting
Middle and Early Wisconsinan	Prairie alluvium
Early Wisconsinan	valley cutting
Sangamon and Late Illinoian	Montgomery alluvium
Illinoian	valley cutting
Yarmouth and Late Kansan	Bentley alluvium
Kansan	valley cutting
Aftonian and Late Nebraskan	Williana alluvium
Nebraskan	valley cutting

On the other hand, those who have worked south from the glacial section correlate the Williana, Bentley, and Montgomery alluvial fills with pre-Nebraskan erosion surfaces (Leighton and Willman, 1950).

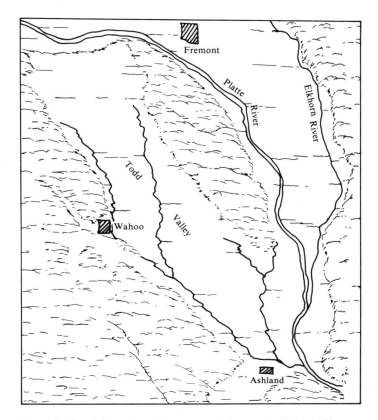

Figure 7-15. Prior to the Nebraskan and Kansan glaciations (Chapter 5), drainage in southern Nebraska was eastward. Those glaciations dammed that drainage, and the present courses of the river developed later. During the Early Wisconsinan glaciation the Platte flowed southeast through Todd Valley, which had been eroded into Nebraskan and Kansas till. During the Wisconsinan, the Platte turned eastward to join the Elkhorn, and the new valley is about 75 feet (23 m) lower than Todd Valley. Width of view, 30 miles (48.3 km) Omaha is 30 miles (48.3 km) east of Wahoo; Lincoln is 30 miles (48.3 km) south. (See Lugn, 1935)

STREAM DEPOSITS IN THE ATLANTIC COAST STATES: POTOMAC AND SUSQUEHANNA RIVERS

The earliest record of the Potomac River is a gravel deposit (Brandywine Formation) on uplands 15 miles (24.1 km) southeast of Washington, D.C. The sandy and gravelly deposits are 200 to 250 feet (61 to 76.2 m) above the present river, overlie Miocene formations, and may be Pliocene or earliest Pleistocene.

On the Piedmont Plateau at Great Falls, 12 miles (19.3 km) upstream from Washington, the Potomac River crosses a series of steeply dipping, resistant veins of quartz. Down-cutting progresses slowly (Fig. 7-16*B*). Falls of that kind are slowly lowered but do not retreat upstream, unlike falls off resistant horizontal ledges (e.g., Niagara Falls), which retreat upstream without being lowered.

Downstream from Great Falls, the Potomac River crosses less resistant metamorphic rocks (schist) and has cut a gorge in them (Fig. 7-16*C*). Below the

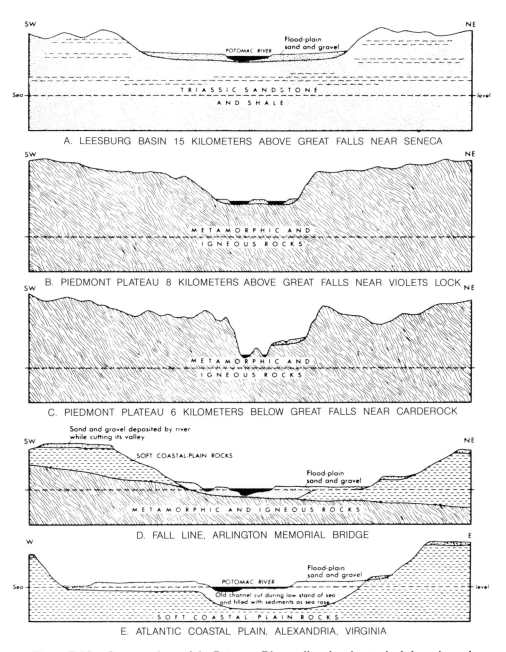

Figure 7-16. Cross sections of the Potomac River valley showing typical shape in each landscape province. Each cross section is approximately 4 miles (6.4 km) long. (From U.S. Geological Survey and National Park Service, 1970)

falls the river has cut a deep, wide valley in the easily eroded coastal plain formations (Fig. 7-16*D, E*); the erosion and fill in the valley extend below sea level because, when that part of the valley was cut, sea level was lower due to glaciations. The fill probably consisted of alluvium and estuarine deposits that were laid down when the sea level rose during the interglaciations. The terraces along the side of the river record successive stages in the down-cutting.

The Susquehanna River originally may have joined the Delaware River on the peninsula east of the Chesapeake Bay. Near Salisbury, Maryland, wells drilled for water encountered a buried river gravel 200 feet (61 m) below the surface. Perhaps the ancestral Susquehanna turned east to the Atlantic. If so, it was captured and turned south by a tributary of the Potomac River. More drilling is needed to reconstruct the old stream courses.

A series of alluvial terraces border the Susquehanna River from its mouth headward to the southern limits of the glacial drift in northern Pennsylvania and southern New York, and the terraces have been correlated with known Wisconsinan drifts and with two or more pre-Wisconsinan drifts. Wisconsinan terraces are equated with the Olean, Binghamton, Valley Heads, and Mankato glaciations in northern Pennsylvania and New York.

REFERENCES CITED AND ADDITIONAL BIBLIOGRAPHY

Albritton, C. C., Jr., and K. Bryan, 1939, Quaternary stratigraphy in the Davis Mountains, Trans-Pecos, Texas, *Geol. Soc. America Bull.* **50:**1423-1474.

Bryan, K., 1923, Erosion and sedimentation in the Papago Country, Arizona, *U.S. Geol. Survey Bull. 730-B*, p. 19-90.

Bryan, K., 1935, *The formation of pediments*, 16th Internatl. Geol. Congress Report, Washington, D.C., pp. 765-775.

Bryan, K., 1954, The geology of Chaco Canyon, New Mexico, *Smithsonian Misc. Colln.* vol. 122, no. 7, 65p.

Bull, W. B., 1964a, Geomorphology of segmented alluvial fans in western Fresno Co., California, *U.S. Geol. Survey Prof. Paper 352-E*, pp. 89-129.

Bull, W. B., 1964b, Alluvial fans and near-surface subsidence in western Fresno Co., California, *U.S. Geol. Survey Prof. Paper 437-A*, 71p.

Condra, G. E., and others, 1947, Correlation of the Pleistocene deposits of Nebraska, *Nebraska Geol. Survey Bull 15*, 73p.

Davis, W. M., 1889, The rivers and valleys of Pennsylvania, *Nat. Geog. Mag.* **1:**183-253.

Denny, C. S., 1956, Surficial geology and geomorphology of Potter County, Pa., *U.S. Geol. Survey Prof. Paper 288*, 72p.

Denny, C. S., 1965, Alluvial fans in the Death Valley region, California and Nevada, *U.S. Geol. Survey Prof. Paper 466*, 62p.

Everitt, B. L., 1968, Use of cottonwood in an investigation of the recent history of a floodplain, *Am. Jour. Sci.* **266:**417-439.

Fairchild, H. L., 1925, The Susquehanna River in New York and evolution of western New York drainage, *New York State Mus. Bull. 256*, 99p.

Fisk, H. N., 1944, Geological investigation of the alluvial valley of the lower Mississippi River Comm., War Dept., Corps of Engineers, 63p.

Fisk, H. N., and E. MacFarlan, Jr., 1955, Late Quaternary deltaic deposits of the Mississippi River (La.)—Local sedimentation and basin tectonics, in *Crust of the Earth*, A Symposium, A. Poldervaart, ed., Geol. Soc. America Spec. Paper 62, pp. 279-302.

Flint, R. F., 1949, Pleistocene drainage diversions in South Dakota, *Geogr. Annaler,* **31:**56-74.

Frye, J. C., and A. B. Leonard, 1952, Pleistocene geology of Kansas, *Kansas State Geol. Survey Bull. 99*, 230p.

Gilbert, G. K., 1877, Report on the geology of the Henry mountains, Utah, U.S. Geographical and Geological Survey Rocky Mountain region (Powell, Survey, Washington), 170pp.

Gilbert, G. K., 1914, The transportation of debris by running water, *U.S. Geol. Survey Prof. paper 86*, 263p.

Gilbert, G. K., 1917, Hydraulic-mining debris in the Sierra Nevada, *U.S. Geol. Survey Prof. Paper 105*, 54p.

Gile, L. H., J. W. Hawley, and R. B. Grossman, 1970, *Distribution and Genesis of Soils and Geomorphic Surfaces in a Desert Region of Southern New Mexico*, Guidebook, Soil Science Society of America, in cooperation with the Department of Agronomy, New Mexico State University, Las Cruces, N.M., 156p.

Gilluly, J., 1937, Physiography of the Ajo Region, Arizona, *Geol. Soc. America Bull.* **48:**323-348.

Graf, W. L., 1983, Downstream changes in stream power in the Henry mountains, Utah, *Assoc. Am. Geographers Annals* **73**(3):373-387.

Hack, J. T., 1942, The changing environment of the Hopi Indians of Arizona, *Harvard Univ. Peabody Am. Archeol. Ethnol. Papers* **35**(1) 85p.

Hunt, C. B., 1954, Pleistocene and Recent deposits in the Denver area, Colorado, *U.S. Geol. Survey Bull. 996-C*, 49p.

Hunt, C. B., 1955, Recent geology of Cane Wash, Monument valley, Ariz., *Science* **122:**583-585.

Hunt, C. B., and others, 1953, Geology and geography of the Henry mountains region, Utah, *U.S. Geol. Survey Prof. Paper 228*, 234p.

Hunt, C. B., and D. R. Mabey, 1966, Stratigraphy and structure of Death Valley, Calif., *U.S. Geol. Survey Prof. Paper 494-A*, 162p.

Johnson, D. W., 1932a, Rock fans of arid regions, *Am. Jour. Sci.*, ser. 5, **23:**389-416.

Kempton, J. P., and R. P. Goldthwaite, 1959, Glacial outwash terraces of the Hocking and Scioto River valleys, Ohio, *Ohio Jour. Sci.* **59:**135-151.

Krieger, A. D., 1950, A suggested general sequence in North American projectile points, *Utah Univ. Anthropol. Papers No. 11*, pp. 117-124.

Leighton, M. M., and H. B. Willman, 1950, Loess formations of the Mississippi valley, *Jour. Geology* **58:**599-623.

Leopold, L. B., M. G. Wolman, and J. P. Miller, 1964, *Fluvial Processes in Geomorphology*, W. H. Freeman and Co., New York, 522p.

Leopold, L. B., and J. P. Miller, A postglacial chronology for some alluvial valleys in Wyoming, *U.S. Geol. Survey Water-Supply Paper 1261*, 90p.

Lugn, A. L., 1935, The Pleistocene geology of Nebraska, *Nebraska Geol. Survey Bull. 10*, 213p.

Malde, H. E., 1971, History of Snake River Canyon indicated by revised stratigraphy of Snake River Group near Hagerman and King Hill, Idaho, *U.S. Geol. Survey Prof. Paper 644-F*, 21p.

Malde, H. E., and H. A. Powers, 1962, Upper Cenozoic stratigraphy of western Snake River Plain, Idaho, *Geol. Soc. America Bull.* **73**(10):1197-1219.

Matthes, F. E., 1930, Geologic history of Yosemite Valley, *U.S. Geol. Survey Prof. Paper 160*, 137p.

Norris, S. E., and G. W. White, 1961, Hydrologic significance of buried valleys in glacial drift, *U.S. Geol. Survey Prof. Paper 424-B*, p. B34-B35.

Paige, S., 1912, Rock-cut surfaces of the desert ranges, *Jour. Geol.* **20:**444.

Peltier, L. C., 1949, Pleistocene terraces of the Susquehanna River, Pennsylvania, *Pennsylvania Geol. Survey Bull. G23*, 4th ser., 158p.

Plummer, F. B., 1932, Cenozoic systems, in *The Geology of Texas*, vol. 1, Stratigraphy, Univ. Texas Bull. 3232 (1966, 5th printing).

Powell, J. W., 1875, *Exploration of the Colorado River of the West and Its Tributaries*, U.S. Govt. printing Office, Washington, D.C., 291p.

Rachocki, A. H., 1961, *Alluvial Fans*, Wiley, New York, 161p.

Rich, J. L., 1935, Origin and evolution of rock fans and pediments, *Geol. Soc. America Bull.* **46:**999-1024.

Schwartz, G. M., and G. A. Theil, 1954, *Minnesota's Rocks and Waters*, University Minnesota Press, Minneapolis, Minn. 366p.

Scott, G. R., 1963, Quaternary geology and geomorphology of the Kassler Quadrangle, Colorado, *U.S. Geol. Survey Prof. Paper 421-A*, 70p.

Sigafoos, R. S., 1964, Botanical evidence of floods and flood-plain deposition, *U.S. Geol. Survey Prof. Paper 485-A*, pp. 1-35.

Stout, W., and D. Shaaf, 1931, Minford silts of southern Ohio, *Geol. Soc. America Bull.* **42:**663-672.

Tight, W. G., 1903, Drainage modifications in southeastern Ohio and adjacent parts of West Virginia and Kentucky, *U.S. Geol. Survey Prof. Paper 13*, 111p.

Trowbridge, A. C., 1954, Mississippi River and Gulf Coast terraces as related to Pleistocene history—A problem, *Geol. Soc. America Bull.* **65:**793-812.

Tuan, Yi-Fu, 1959, Pediments in southeastern Arizona, *Calif. Univ. Pubs. Geography*, vol. 13, 163p.

U.S. Geological Survey and National Park Service, 1970, *The geological history of Great Falls and the Potomac River gorge*, Misc. Pubs., Washington, D.C., 46p.

Wolman, M. G., Fluvial Processes, in *Encyclopedia Britannica*, vol. 7, pp. 437-446.

EOLIAN DEPOSITS

EXTENT AND KINDS

Every physiographic province has its eolian deposits. These include sand in various kinds of dunes, especially along beaches and in scattered sand sheets and dunes in deserts, all favorite playgrounds. More extensive and more productive than the sand deposits, however, are the silty deposits of windborne loess that cover much or most of the agricultural heart of America in the central United States. Another, less desirable kind of eolian deposit includes aerosols and particulates contributed by automobile exhausts and by stacks at manufacturing plants, smelters, and refineries. In industrial cities even high-level apartments daily collect a smudge of stack particulates on the sills and along the street at rush hours, and in the canyons between the high rises, brownish smog created by congested automobile traffic commonly forms a layer 50 feet (15 m) high. Although these urban effluents are not shown on the map, they are of tremendous environmental importance, partly because of health hazards and partly because of dirt in and on buildings and clothing. Hillsides may be defoliated and even deforested. Another kind of wind-blown deposit, also not shown on the map but a definite nuisance, is the accumulation of wind-blown litter along fence lines.

Wind-blown dust creates a desert haze in the arid and semarid west. On clear days, like after a rainstorm, one may be able to see mountains a hundred miles away. At night, one can see sixth magnitude stars, a gorgeous panorama of stellar geography. When there is haze, even nearby mountains may become hidden, and nights reveal no better than third or fourth magnitude stars. The conditions of course are made worse by emissions from urban areas, automobiles, and power plants, and by dust from unpaved roads. How to minimize the air pollution is a major political issue where parks and other scenic areas are threatened by overuse.

Volcanic ash is an important constituent of eolian silts in the western United States and contributes to the high content of swelling clays in many of the fine-grained western deposits, including those reworked by streams or deposited in playas. Also important in western eolian deposits, especially in the Southwest, is wind-blown calcium carbonate, which contributes to the development of caliche there (Fig. 2-6). This dust is also a nuisance because dusty windows spattered by light rain may need to be cleaned with vinegar or other acid. The need for such housecleaning effort emphasizes the importance of wind-blown dust in the origin of caliche.

Locally on shale deserts, where badlands have developed, protracted drought may cause flaking of the shale, and wind collects the flakes in shale dunes. These unusual features are, of course, ephemeral, because the next rain reduces the dune to mud. Still another kind of eolian deposit develops where wind sweeps across saline flats, eroding salt efflorescences and depositing them to the leeward of the saltpan. Gravel fans on the leeward (eastern) side of Death Valley are notably more saline than those on the windward (western) side. Gypsiferous dunes have developed on the lee side of the Great Salt Lake desert, and similar dunes have formed at White Sands National Monument.

The map shows sand sheets (s), mostly with sand dunes, on the Great Plains. These are sand sheets over Pleistocene fan deposits (s/osg), and certain loess deposits: Wisconsinan loess (wl), which is little weathered, and pre-Wisconsinan loess (es), which is deeply weathered.

Wind Transport and Deposition

Wind transport of dry sand and silt is controlled partly by wind velocity, partly by the quantity and coarseness of the source materials, partly by the extent of bare ground, and partly by the roughness of the ground surface, especially by the kind and scatter of vegetation. Transportation is by direct impact of the wind and by frictional drag along the sides of the particles, which move by rolling, saltation, or suspension. Generally there has been found a linear relationship between wind velocity and sizes of particles being transported (Thoulet, 1884; Bagnold, 1941). Wind velocities and sizes of particles moved in suspension are:

Velocity	Diameters of particles moved in suspension
5 meters/sec = 11 mph	
0.5 meters/sec	0.04 mm
1.0 meters/sec	0.08 mm
5.0 meters/sec	0.41 mm
10.0 meters/sec	0.81 mm

Strong winds move coarse sand by dragging or rolling the sand grains along the ground. Above this layer is laminar flow 10 to 30 meters high, and the effect is like a ground blizzard (Fig. 8-1*A*). Where the source material is fine silt, turbulent winds create a dust cloud that may be a kilometer or more high (Fig. 8-1*B*). Very fine dust may be carried higher and transported long distances, even thousands of miles, before falling.

Sand becomes deposited as dunes, which may form climbing dunes in the lee direction. The dunes forms may be barchans (Fig. 8-2*A*), transverse (Fig. 8-2*B*), parabolic (the opposite of barchan, Fig. 8-3*A*), or longitudinal (Fig.

A

B

Figure 8-1. Sand storms. *Top:* Storm of coarse sand resembling a ground blizzard, Green River Desert, Utah. *Bottom:* Dust storm (light area) caused by turbulent winds on silty ground at Organ Pipe Cactus National Monument, southern Arizona.

8-3*A, B*). In flat areas the sand collects as a smooth sheet with only ripples on the surface.

Vegetation collects blowing sand, and the size of the resulting mounds (coppice dunes) depends on the size of the shrubs. Mounds collected at oak brush or greasewood on the Green River Desert, Utah, or at mesquite and saltbush in southern New Mexico may be 10 feet high. Smaller mounds collect around smaller shrubs like blackbrush or Mormon tea. But are the shrubs different because of differences in the amount of sand, or is the amount of sand different because of differences in the size of the shrubs?

A

B

Figure 8-2. Barchan dunes (*top*) and transverse dunes (*bottom*). The barchan dunes are at Quincy Basin south of Grand Coulee in Washington; the transverse dunes are at Organ Pipe Cactus National Monument.

A

B

Figure 8-3. Longitudinal dunes. *Top:* On Green River Desert, Utah, longitudinal dunes form tails back of clustered barchan and parabolic dunes. Wind from lower left. *Bottom:* Ground view of ridge formed by longitudinal dune on the Navajo Reservation in northwest New Mexico. This particular ridge is 20 feet (6.1 m) high, about 100 feet wide (30.5 m), and 2 miles (3.2 km) long.

Figure 8-4. Erosion features in massive sandstone on the Colorado Plateau, showing the combined effects of wind and of water (see also Fig. 8-5). Tanks are shallow horizontal depressions where water collects at the surface and dissolves the cement holding the sand grains in the rock, which then allows wind to remove the loosened grains. (Photograph by Arthur Hohl, Johns Hopkins University)

SAND

Sources: Wind Erosion

The ultimate sources of the sand are weathered sandstone formations, weathered granitic rocks, or beach sands. Massive sandstone is pitted with depressions (tanks) and niches (Figs. 8-4, 8-5), which are places where the calcium carbonate cement is weak or there is water seepage. Water seeping through such rocks dissolves the cement, and the loosened sand grains can be blown away. Niches tend to be aligned along bedding planes. Arches, like those at Arches National Park, are also mostly due to solution; they are not wind-erosion features because the wind has served only to remove the loosened grains. The natural bridges at Bridges National Monument and Rainbow Bridge, which resemble the arches, are the result of lateral erosion at stream meanders.

Related to the niches and arches are the big alcoves, overhangs in the sandstone formations. Many on the Colorado Plateau were used by the prehistoric Pueblo Indians for protecting dwellings like those at Mesa Verde (Fig. 8-6), Betatakin, Canyonlands, and Canyon de Chelly. Most of these big overhangs formed where springs discharge from the base of the sandstone; the seepage dissolves the cement, and the wind removes the loose sand. The overhangs provided not only shelter but also a built-in water supply, further emphasizing that the process is not wind erosion but chiefly solution followed by wind transport.

Figure 8-5. Niches in steep or even vertical rock faces are formed in much the same way as are tanks. Solution of the cement holding the sand grains together is favored along layers where the cement is incomplete between the grains.

Other characteristic features of sand deserts in the southwestern United States are monoliths left by erosion (Fig. 8-7). Some are pedestal rocks like Mexican Hat in southern Utah and Camel Rock near Santa Fe, New Mexico.

In the Southwest, plowed fields are a major source of wind-blown sand and dust, the condition that prevailed in the dust-bowl areas of the southern Great Plains during the 1930s. Highways in southern Arizona have signs reading "Caution—blowing sand; turn on headlights." This in a sunny desert! Ravished, wind-eroded fields after a drought may have sand heaped as high as fence posts along the lee side of the plowed area. During the dust-bowl days, the geology department at the University of Wichita estimated, during one dust storm, that the atmosphere to a height of 1 mile (1.6 km) over the 30 square miles (77.7 square km) of Wichita contained 5 million tons of dust (Lugn, 1935, p. 166). Some of the dust fell as far east as Washington, D.C., enough to cause this author to rinse his cars! However, most such dust travels only short distances and is in effect replacing topsoil lost previously on lands downwind. At many places on the southern Great Plains, wind erosion is clearly indicated by sand mounded on the lee (eastern) sides of depressions, the so-called buffalo wallows.

Although the principal role of wind is to transport sand to build the constructional forms already described, it can erode by sandblasting (corrasion). This is accomplished mostly by sand moving in a layer within about a foot of the ground. The bases of telephone poles may be so abraded that they need to be protected by stones placed around them. Above that the sand and dust move without much corrasion. Nevertheless, telephone wires may be so worn they must be replaced.

Sandblasting produces facets on rock surfaces. The faceted surfaces may

Figure 8-6. Prehistoric Anasazi cliff dwelling under an overhanging cliff at Mesa Verde National Park. Such sheltered alcoves were widely used for dwellings during the period A.D. 500 to A.D. 1300. Springs issuing from the base of the cliffs forming sandstone dissolved the sandstone's cement, loosening the sand grains, which could be blown away by the wind. The alcoves used for dwellings had a built-in water supply. Streaks of desert varnish antedate the ruins.

Figure 8-7. Monoliths on the Colorado Plateau illustrate the relative roles of water and wind in desert erosion. Erosion of the monoliths has been by water dissolving the calcium carbonate cementing sand grains in the rocks. The loosened sand grains fall and blow away so debris does not accumulate around the bases of the monoliths. Brigham Butte, at the west edge of the Green River Desert about 6 miles north of Hanksville, Utah, is a monolith of Jurassic sandstone.

be smooth, pitted, or fluted. Where the flutings come together like braids, the rocks (especially limestone) are referred to as rillenstein and are attributed to wind-driven rainwater. Similar surfaces on massive rock salt, clearly due to wind-driven rainwater, support the interpretation. Recently discarded glass bottles become frosted, and a windshield can become frosted, too, if the car is driven against a sand storm.

Stratigraphy of Sand Dunes

An example of the stratigraphy at dunes on the barrier beach at Cape Hatteras has already been described (Fig. 4-10). Other examples are dunes along the Atlantic Coast that overlap Pleistocene shell beds. In the western deserts the stratigraphy is similar to that found in alluvial and colluvial deposits, with erosional unconformities and weathered zones separating younger from older deposits.

Most or all the active dunes in the southwest are found where old stabilized dunes are being reworked. Figure 8-8 illustrates the kind of stratigraphic relations that can be seen.

In the Southwest this general stratigraphy is further reinforced by the physical relationships between dunes and alluvial deposits that can be dated paleontologically and archaeologically. The older dunes, for example, extend across Late Pleistocene alluvium and fall into arroyos eroded into that old alluvium. The dune sand in the arroyo in turn may be overlapped by the Middle Holocene alluvium, identified by its modern fauna and prepottery lithic artifacts. And the modern dune overlies that Middle Holocene alluvium and the pottery sites built on it.

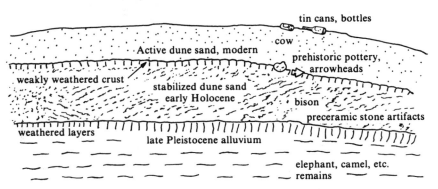

Figure 8-8. Idealized diagram illustrating the kind of stratigraphy that can be found in sand dunes on the Great Plains and Colorado Plateau. Early Holocene dune sand is weathered, dunes are stabilized, and contain artifacts of pre-ceramic Indian occupations associated with modern mammalian remains such as *Bison bison*. Beneath this is late Pleistocene alluvium with weathered surface layers and containing remains of Pleistocene animals such as elephant and camel. On top of the stabilized sand are pottery, arrowheads, and other remains of the ceramic Indian occupations. Surface layers of the active sand have tin cans, bottles, and other historical archeological objects. Based on Hunt, 1954 and Hunt, 1955.

Distribution and Extent of Dunes

Dunes have developed wherever the ground is sandy. They are therefore coextensive with the sandy beaches along the coasts and around the Great Lakes, and they are developed on sandy areas of the coastal plain (rs) inland from the Atlantic and Gulf coasts—for example, in New Jersey and along the inner edge of the coastal plain in the Carolinas (see page 22).

Along the one hundredth meridian and farther west, the map shows the most extensive sand sheets, all of which typically have dunes. Other areas having extensive, well-developed dunes include much of the Wyoming basins and plateau surfaces on the Colorado Plateau. In the Great Basin dunes are extensive on the lower parts of the sandy fans (fs). Many others are developed on the sandy toes of the gravel fans (fg) and on the lee (eastern) side of Pleistocene lake deposits (|). Two notable exceptions to the generalizations (neither shown on the map) are: (*a*) in Death Valley, where dunes around the saltpan are of limited extent because the sand there is securely held down by salts; and (*b*) at Coral Sands State Park in southern Utah east of Zion National Park, where dunes have drifted northeastward onto shaley ground (sh) north of the bedrock cliffs (br) along the Arizona border.

These areas with dunes aggregate almost 150,000 square miles (388.500 sq. km), nearly 5 percent of the conterminous United States. They are, moreover, second only to water surfaces as popular playgrounds. Many state parks are located at dunes, and two fields of dunes have been set aside as part of the nation's national parks. One of these, Great Sand Dunes National Monument, is in the San Luis Valley of southern Colorado. The other, White Sands National Monument, is in southern New Mexico.

Great Sand Dunes National Monument covers about 150 square miles (388.5 km). The largest and most spectacular dunes there are transverse, but there are also large areas of climbing dunes, longitudinal dunes, and parabolic dunes. The dune field, derived from the alluvial fan built by the Rio Grande where it discharges from the San Juan Mountains, is along the east side of the San Luis Valley, and the dunes are climbing the western slope of the Sangre de Cristo Range.

White Sands National Monument in the Tularosa Valley in southern New Mexico covers about 300 square miles (777 sq. km). The dunes there are unusual in being composed of gypsum. That part of the Tularosa Valley contained a Pleistocene lake, Lake Otero, which also covered about 300 square miles (777 sq. km) but centered somewhat southwest of the dunes. The lake became saturated with gypsum, in part perhaps brought up by capillarity of shallow groundwater and in part washed from gypsum-bearing Permian formations in the adjoining mountains. The modern lake, known as Lake Lucero, is dry playa most of the time but is flooded in wet periods. Winds from the southwest transport fragments of the gypsum from the edges of the playa, and the fragments (hardness only 2) quickly become rounded grains. The 1/7,500,000 map shows the geographic position of the sand(s) is northeast of Lake Lucero (1). At the lake, solitary dome-shaped dunes are developed; farther northeast are transverse dunes that gradually merge into a field of barchans. Near the margin of the dune area, where there is more vegetation, the dunes are parabolic. Dunes of gypsum (not shown on the map) have also developed along the east edge of the Great Salt Lake desert.

In the Great Salt Lake area, near the Bingham Canyon copper mine, the

Utah State Parks Commission has set aside an area of sandy, abandoned mill tailings. The tailings have been reworked into low dunes, and the commission set the area aside as an obstacle course for off-road vehicles, an example of waste land salvaged to provide a much-used playground.

Other major areas of dunes in the United States besides those at the sandy beaches already described (Chapter 4) are:

1. Those along sandy formations (rs) of the Atlantic Coastal Plain (Fig. 1-5);

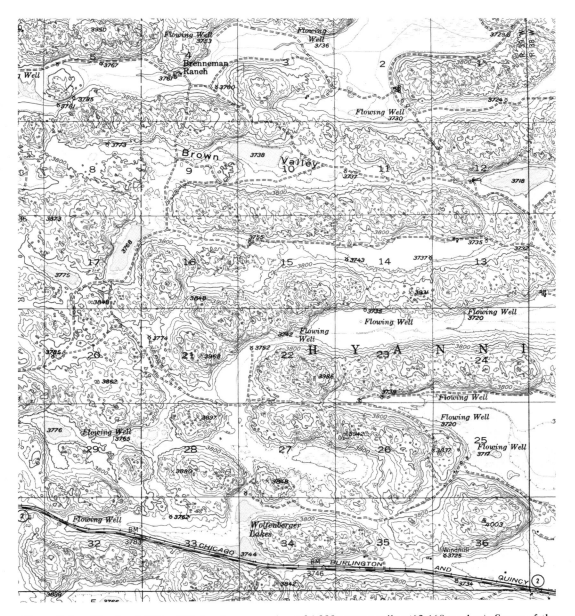

Figure 8-9. The Sand Hills of Nebraska cover about 24,000 square miles (62,160 sq. km). Some of the dunes are a few hundred feet (100–200 m) high. Between the sand hills are meadows, some of which contain ponds. (From U.S. Geological Survey Ashby 15-minute quadrangle map. 1948)

2. Those around the Great Lakes, especially along the sandy Pleistocene shorelines of the ancestral lakes (not shown on the map);
3. The Sand Hills (s) of northwestern Nebraska (Fig. 8-9), derived from Tertiary formations farther west and forming one of the most extensive sand areas in North America;
4. A similar but much smaller set of sand hills (s) in northeastern Colorado probably also derived from sandy Tertiary formations but perhaps added to by sand blown from the South Platte River when it was flooded by glacial meltwaters;
5. The Navajo country in northeastern Arizona and northwestern New Mexico (ss);
6. Scattered areas (not shown on the map) in southern Arizona (Fig. 8-2B) and southern New Mexico (Fig. 8-4);
7. Coral Pink Sand Dune State Reserve, Utah (not shown on the map), 10 miles (16.1 km) west of Kanab;
8. Little Sahara, north of Delta, Utah (not shown on the map);
9. The Green River Desert (Fig. 8-3) and Burr Desert (ss) east of the San Rafael Swell and Henry Mountains, Utah;
10. The Red Desert in Wyoming (ss), 50 miles (80 km) east of Rock Springs;
11. Kilpecker dune field, Wyoming (ss), 30 miles (48 km) north of Rock Springs;
12. Great Salt Lake desert, Utah (not shown on the map);
13. A field north of the Humboldt River in north-central Nevada (not shown on the map); and
14. Another (not shown) in the Quincy Basin south of Grand Coulee in Washington (Fig. 8-2A);
15. A belt (not shown) extending west from the Colorado River to the Imperial Valley in southern California, perhaps derived from the alluvial fan of the Colorado River; and (also not shown)
16. In southeastern California and southern Nevada, the Mojave and Amargosa deserts and Death Valley.

LOESS

The 1/7,500,000 map shows the great extent of loess (es and wl) in the central United States. It buries most of the pre-Wisconsinan glacial deposits. At times of thaw during the Pleistocene, when the glaciers melted, the rivers discharged vast floods, and as the floodwaters receded, the bottoms were left as bare mud flats. Prevailing winds from the west whipped up clouds of dust and deposited the silt as loess on uplands to the east. The map shows a belt of loess (es) along the east bank of the Mississippi River. The loess thins and becomes finer grained eastward from the rivers (Fig. 8-10). Very little was carried westward.

Interbedded with the loess are water-laid layers of gravel or sand evidently recording rainstorms and washings that interrupted the accumulation of the silt. Loessial deposits are among our most fertile kinds of ground, but they also are highly subject to erosion. Nonetheless, the loess stands in vertical cliffs because it is massive, fine grained, and divided by vertical columnar joints.

In Nebraska the principal sources of loess were the alluvial lands, sand hills, old till sheets, badlands, silty Tertiary formations, and the High Plains

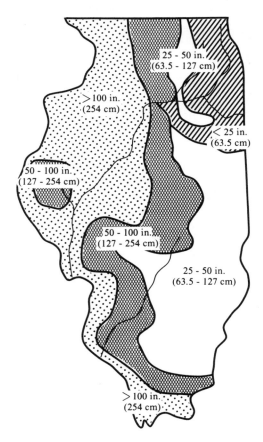

Figure 8-10. Approximate thickness of loess on uneroded topography in Illinois. Note that the loess thins eastward from the Mississippi River, the bed of which was its source. (After Smith, 1942)

areas. Three main loess formations there are the Loveland (Sangamon inter-glaciation), Peorian (mid-Wisconsinan interglaciation), and Bignell loess, which is Late Wisconsinan or Holocene (Condra and others, 1950, pp. 42-46). The appearance and general stratigraphy of the middle western loesses are illustrated in Figure 8-11 *(top)*.

Loess and other eolian deposits are extensive in parts of the plateaus and basins west of the Rocky Mountains, and some of these deposits include volcanic ash. A deposit of pre-Wisconsinan loess in the eastern part of the Colorado Plateau, southeastern Utah, and southwestern Colorado (es on the map; Fig. 8-11, *bottom*) is unusual in the United States in having a desert source. It is probably the fine-grained equivalent of the sand dune fields on the Navajo reservation in northeast Arizona. This deposit, therefore, is more like those of the Near East or Asia than like those of the central United States. Another extensive deposit of loess (wl on the map) blankets the lava plateaus of eastern Washington and Oregon. It is referred to as Palouse soil, and most of it is pre-Wisconsinan, although blanketed by younger loess. Its Pleistocene age is indicated by several occurrences of mammoth remains. Other loess overlies lavas on the Snake River plain in southern Idaho (es on the map).

Figure 8-11. Loess. *Top:* Two ages of loess along the Ohio River where Wisconsinan loess overlies pre-Wisconsinan loess. Geologist points to the contact. *Bottom:* Pre-Wisconsinan loess in southeastern Utah has a strong layer of caliche.

REFERENCES CITED AND ADDITIONAL BIBLIOGRAPHY

Bagnold, R. A., 1941, *The Physics of Blown Sand and Desert Dunes*, William Morrow, New York.

Condra, G. E., and others, 1950, Correlation of the Pleistocene deposits of Nebraska, *Nebraska Geol. Survey Bull. 15-A*, 74p.

Cooper, W. S., 1958, Coastal Sand Dunes of Oregon and Washington, *Geol. Soc. America Mem. 72*, 169p.

Dodge, N. N., 1971, *The Natural History Story of White Sands National Monument*, Southwest Parks and Monuments Association, Globe, Ariz., 69p.

Flint, R. F., 1955, Pleistocene Geology of Eastern South Dakota, *U.S. Geol. Survey Prof. Paper 262*, 173p.

Frye, J. C., and A. B. Leonard, 1952, Pleistocene geology of Kansas, *Kansas State Geol. Survey Bull. 99*, 230p.

Grabau, A. W., 1924, Eolian deposits, in *Principles of Stratigraphy*, reprinted by Dover, New York, pp. 549-578.

Hack, J. T., 1941, Dunes of the western Navajo country, *Geog. Rev.* **31**:240-263.

Hack, J. T., 1942, The changing environment of the Hopi Indians of Arizona, *Harvard Univ. Peabody Mus. Am. Archeol. Ethnol. Papers*, vol. 35, 85p.

Huffington, R. E., and C. C. Albritton, Jr., 1941, Quaternary sands on the southern High Plains of western Texas, *Am. Jour. Sci.* **239**:325-338.

Hunt, C. B., 1954, Pleistocene and Recent deposits in the Denver area, Colorado, *U.S. Geol. Survey Bull. 996-C*, 49p.

Hunt, C. B., 1955, Recent geology of Cane Wash, Monument Valley, Arizona, *Science* **122**(3170):583-585.

Hunt, C. B., and D. R. Mabey, 1966, Stratigraphy and Structure, Death Valley, California, *U.S. Geol. Survey Prof. Paper 494-A*, 162p.

Johnson, R. B., 1967, The great sand dunes of southern Colorado, in Geol. Survey Research 1967, Chap. C, *U.S. Geol. Survey Prof. Paper 575-C*, pp. 177-183.

Leighton, M. M., 1950, Loess formations of the Mississippi Valley, *Jour. Geology* **58**:599-623.

Leonard, A. B., and J. C. Frye, 1956, Quaternary (Pleistocene) stratigraphy in central United States of America, IV Cong. Internat. du Quaternaire, Rome-Pise, 1953. Actes [v.] 2, Rome, p. 877-884.

Lugn, A. L., 1935, The Pleistocene geology of Nebraska, *Nebraska Geol. Survey Bull. 10*, 223p.

Lustig, L. K., and D. W. Goodall, 1974, Deserts, in *Encyclopedia Britannica*, vol. 5, pp. 602-620.

Norris, R. M., 1966, Barchan dunes of Imperial Valley, California, *Jour. Geology* **74**:292-306.

Norris, R. M., and K. S. Norris, 1961, Algodones dunes of southeastern California, *Geol. Soc. America Bull.* **72**:605-619.

Olson, J. S., 1958*a*, Wind-velocity profiles, Pt. I of Lake Michigan dune development, *Jour. Geology* **66**:254-263.

Olson, J. S., 1958*b*, Lake-level, beach, and dune oscillations, Pt. 3 of Lake Michigan dune development, *Jour. Geology* **66**:473-483.

Price, W. A., 1958, Sedimentology and Quaternary geomorphology of south Texas, *Gulf Coast Assoc. Geol. Socs. Trans.* **8**:41-75.

Reeves, C. C., Jr., 1965, Chronology of west Texas pluvial lake dunes, *Jour. Geology* **73**:504-508.

Schultz, C. B., and others, 1951, A graphic resume of the Pleistocene of Nebraska (with notes on the fossil mammalian remains), *Nebraska State Mus. Bull.*, vol. 3, n. 6, 41p.

Sharp, R. P., 1963, Wind ripples, *Jour. Geology* **71**:617-636.

Smith, G. D., 1942, Illinois loess—variations in its properties and distribution, *Illinois Univ. Agr. Expt. Sta. Bull. 490.*

Smith, H. T. U., 1950, Physical effects of Pleistocene climate changes in non-glaciated areas; eolian phenomena, frost action, and stream terracing, *Geol. Soc. America Bull.* **60:**1485-1515.

Smith, H. T. U., 1965, Dune morphology and chronology in central and western Nebraska, *Jour. Geol.* **73:**557-578.

Smith, H. T. U., and C. Messinger, 1959, *Geomorphic studies of the Provincetown dunes, Cape Cod, Massachusetts,* Amherst Geology Dept., University of Massachusetts, June, 62p.

Stuntz, S. C., and E. E. Free, 1911, Bibliography of eolian geology, *U.S. Dept. Agriculture, Bur. Soils Bull. 68.*

Thoulet, J., 1884, Experiences relatives a la vitesse des courants d'eau or d'air susceptibles de maintenir en suspension des grains mineraux, *Ann. Mines* **5:**507-530.

PART
IV

MISCELLANEOUS DEPOSITS

BASALT, BEDROCK, CLINKER, AND HOT SPRING DEPOSITS

BASALT

Basaltic lavas (b) shown on the map include some slightly less basic lavas that properly are classified as andesites. The lavas are resistant rocks that greatly control both the landforms and the kind of ground (the surficial deposits) on and about them. Moreover, because the lavas are young, they are either bare rock or only shallowly covered by surficial deposits. The lavas therefore have an important environmental impact because there is great difference in the potential uses of, for example, sandy ground shallowly underlain by basaltic lava and similar ground underlain by other surficial deposits such as alluvium.

In the United States extensive areas of basaltic lava extend from Yellowstone Park westward along the Snake River plains and on the plateaus of eastern Washington and Oregon. In much of this area the lavas are overlain by loess (wl/b), far more than shown on the map. The loessial deposits are thin and discontinuous, but they provide ground favorable for agriculture. Other extensive areas of basalt are on the southwestern part of the Colorado Plateau, along the Rio Grande Valley north of Santa Fe, and in northeastern New Mexico. Small areas of lavas, few of which are shown on the map, are scattered through the Basin and Range Province (Fig. 9-1) and along the Sierra Nevada.

The youngest lavas, located in present valley bottoms, can be dated archaeologically and are no more than a few hundred years old. These young Holocene lavas are mostly bare rock with rough surfaces caused by the freezing of a crust over the molten lava, the crust being dragged by the underflow to form pressure ridges. In many places the crust collapsed where the river of molten lava drained from under it. In the Southwest, such ground is known as malpais (Spanish for "bad ground"), and the lavas are cavernous. Some have caves containing perennial ice. Vegetation on the youngest flows is scanty except for lichens. Flowering plants grow only where the flows are old enough to have collected sand and dust in depressions and cracks. These young lavas are not at all weathered, and many look fresh enough to be still warm.

Some of the basaltic landforms, although too small to show on the map, are

171

Figure 9-1. Pavant Butte, in the volcanic field in the Lake Bonneville basin west of Fillmore, Utah. This is a Late Pleistocene volcano that erupted during the time of Lake Bonneville (Chapter 6), the lake that covered most of northwest Utah. Clearly, the volcanic cone was built before the conspicuous shoreline was eroded on its sides, but there were eruptions after the beach was eroded, as indicated by ash beds on it. The foreground is basaltic lava mantled by lake bottom sediments. (From Gilbert, 1890)

spectacular. Lava-capped mesas are surmounted by lava or cinder cones (Fig. 9-2, *top*). Erosion at the edges of the mesas at cones has produced natural cross sections of volcanoes (Fig. 9-2, *middle*). Further erosion has left the volcanic pipe isolated as a volcanic neck (Fig. 9-2, *bottom*).

Cinder cones are favorite sources of light aggregate, porous rocks useful for base courses or road surfacing and as decorative ground around residences or industrial areas where desert landscaping is desired instead of lawn. In Arizona and New Mexico the cinder cones are numerous enough to provide plenty of material in the numerous volcanic fields. However, to reduce haulage, the cones nearest the highways are favored, and the scenery is apt to be spoiled by having the quarry on the side facing the highway.

In eastern Washington the 1/7,500,000 map shows a dendritic pattern of basalt (b) north of the Snake River. This is a highly diagrammatic representation of a unique area known as the channeled scablands. The bare basaltic surfaces— the "scabs"—were exposed when episodic floods (the Spokane Floods) from Lake Missoula removed the loess cover along huge channels, some of which were eroded deeply into the basalt (as at Grand and Moses coulees). Along the basaltic channels are huge erratics of granitic rocks evidently transported in

Figure 9-2. Volcanic landforms in the southeast part of the Colorado Plateau, at the Mount Taylor volcanic field about 45 miles (72 km) west of Albuquerque. *Top:* Smooth, vegetated volcanic cone rises from a lava-capped mesa at the foot of Mount Taylor. In the foreground a mantle of loessial silt or alluvium mantles the lava flows. *Middle:* Natural cross section of a basaltic volcano exposed by erosion. The volcano came up in a cylindrical pipe through Cretaceous and older formations (ledges in middle right) and through overlying Tertiary rhyolitic tuff (white ledge near the top). It built a cone on top of the tuff and spread lavas for miles around. *Bottom:* Cabezon Peak, a volcanic neck similar to the pipe shown in the middle photograph but isolated by complete removal of the lavas that once surrounded this volcano. The ledges around the base of the neck are Cretaceous sandstone. The slopes are mantled by colluvium.

icebergs, huge gravel bars, and ripple marks that also are disproportionately large. The Spokane Floods, like the Lake Bonneville Flood (Fig. 6-7), exemplify catastrophism.

Mount St. Helens erupted in 1980, but it was an explosive eruption, not the quiet effusion of basalt. It erupted silica-rich ash that fell mostly in an oval area extending northeastward across Washington. In 1914-1917 Lassen Peak in northern California spilled lava over the northwest and northeast rims of the crater, and the area still has hot springs and fumeroles. One of the most recent eruptions on the Colorado Plateau was at Sunset Crater in the San Francisco volcanic field, Arizona. The eruption, about A.D. 1075, buried an Indian village with cinders. Another very young flow is along the San Jose River a few miles east of Grants, New Mexico.

Older Pleistocene lavas form benches tens or scores of feet (5 to 30 m) higher than the present streams. Their surfaces are smoother than those on the Holocene lavas because of fine-grained sediment washed or blown onto them. They commonly have moderate stands of shrubs or even woodland. The rocks are fresh and firm but have a yellow oxidized stain a foot or two (0.3 to .61 m) from the top.

The earliest basaltic lavas of the Southwest cap high mesas and plateaus, as at Grand Mesa in Colorado, the Mount Taylor area in New Mexico, and some of Utah's high plateaus. Their surfaces may be mantled with loess and/or alluvium, and most are high enough to be forested. Hillsides below the lava caps are covered by colluvium; landslides are common (Fig. 3-6).

In the Southwest the extensive lava fields are interrupted by small volcanic cones (Fig. 9-2A). In the Pacific Northwest, such cones are numerous in southeastern Oregon and in places along the Snake River plain. They are lacking in northeast Washington; the extensive basaltic lavas there were erupted largely from swarms of dikes. Much of the ground in southeast Oregon is mantled by scoria. The Malheur River, which drains northeast to the Snake River, was dammed by a lava flow that has created Lakes Harney and Malheur.

In brief, all but the youngest basaltic lavas are largely mantled by transported surficial deposits of sufficient depth and extent to be shown on large-scale maps. But the landforms and kinds of deposits on them are distinct from the other deposits shown on the 1/7,500,000 map. The cover of surficial deposits on the lavas is generally stony, with blocks of the lavas or scoriae frost-heaved upward into the finer-grained cover. Even on the oldest surfaces, the presence of underlying lava is usually unmistakable. Not shown on the map are the Triassic basaltic (diabase) sills and lavas in the northeastern states, notably the Watchung Ridges in northern New Jersey, the Palisades along the Hudson River, and the similar ridges along the Connecticut Valley. These areas were glaciated and, except for the steep rocky cliffs too small to show on the map, are covered by glacial deposits. Similar rocks on the Piedmont Plateau south of the glaciated area are weathered to deep residiuum and included in r/Tr. The most easterly lavas are in northeastern New Mexico.

BEDROCK

The map shows bedrock (br) in only two regions—the canyon lands on the Colorado Plateau and ranges in the Basin and Range Province. In these areas, both of which are arid, probably 25 percent or more of the ground is bare rock; in places the bare rock may extend over 60 percent of the ground, and only 40 percent is mantled by thin alluvium, residuum, wind-blown sand, or colluvium. By contrast, even the Rocky Mountains, despite their name, average only about 10 percent bare bedrock, the rest being covered mostly by colluvium but with some alluvium and glacial deposits, and, on flat uplands, thin residuum. These estimates are confirmed by Forest Service land inventories; in the Appalachians, estimates by the U.S. Geological Survey, state surveys, and the Forest Service indicate only about 5 percent bare rock. On the map, those mountainous areas are shown as colluvium (co).

On the Colorado Plateau, the bare rock along the canyons is mostly Jurassic and older sandstone. Ledges are bare except for thin residuum or small alluvial fans. Shale bluffs have discontinuous trains of colluvium in gullies. The laccolithic mountains on the Colorado Plateau like the Rocky Mountains, are largely mantled by colluvium and periglacial and glacial deposits.

In the Basin and Range Province the ranges consist of many different kinds of bare rock. Some are Precambrian metamorphic rocks. Some are Paleozoic marine sedimentary rocks. Some are Late Cretaceous or Early Tertiary granitic rocks. Some are basaltic lavas but are too small in area to show as Qb or QTb on the 1/7,500,000 map. Some ranges consist of uplifted blocks of Tertiary playa sediments and various kinds of poorly consolidated volcanic rocks, mostly rhyolite (the explosively eruptive equivalent of silica-rich granite). These ranges of Tertiary rocks are usually bordered by fan sand (fs) or fan silt (fsi), while the older, consolidated rocks are bordered by fan gravels (fg).

CLINKER (KK)

The map shows spots of clinker on the northern Great Plains, especially in the lignite and subbituminous coalfields of northeastern Wyoming and eastern Montana. Clinker deposits formed where the lignite or coal beds burned at the outcrops, and the burning extended underground back from the outcrop to where there was insufficient oxygen to maintain the fire. The clinkered rock, quite like the clinker in a coal furnace, includes shale or sandstone partings that were in the coal bed and, depending on the heat of combustion, a few to many feet of the roof rocks, which collapsed as the coal burned under its roof.

Most of the fires probably started spontaneously, because the burned areas are more numerous and more extensive in the lignite fields than in the fields of subbituminous coal that contain less volatile matter. Some fires, though, were probably started by lightning or by grass fires.

The map is misleading in showing the clinker occurring in spots. Actually the clinker extends in nearly horizontal bands that conform to the position of the outcrops of the burned beds along the sides of hills and valleys. Not shown on the map are small, scattered areas of clinker at burned coal beds on the Colorado Plateau. Those coal beds are mostly Cretaceous and have less volatile matter than do the Tertiary lignite beds on the northern Great Plains.

The clinker is reddened by oxidation of the iron. In places the oxidation is sufficient to affect a compass needle. The rock is tough, commonly with a glassy sheen, and is widely used for railroad ballast and road metal.

HOT-SPRING DEPOSITS (HS)

Thermal springs are usually considered to include those springs having temperatures about 10° F warmer than the average annual temperature of the locality. Using this definition, the conterminous United States has about 1,200 thermal springs. For the purpose of the 1/7,500,000 map, the cutoff was taken at a water temperature of about 100° F (about 22.5° C). Moreover, because of the scale of the map, at many places one symbol represents clusters of hot springs. Nevertheless, the map shows that hot springs and their potential geothermal energy are almost all located in the geologically youthful western states; Yellowstone National Park is the classic example. Most hot springs are found where there has been Pleistocene volcanism or faulting. The few hot springs in the central and eastern states are attributed to deep circulation of groundwater. In the west, hot rocks are generally shallower and groundwater is correspondingly more readily warmed.

Most of the hot-spring deposits shown were deposited by groundwater circulating along faults or other fractures that extend up from the heated rocks. In some places, part of the water may be contributed by emanations from magmas at depth. A few of the hot springs have large discharges, several thousand gallons per minute, but most discharge no more than a few hundred gallons per minute. Discharges may be steady or intermittant, as at geysers.

The deposits, travertine, are mostly calcium carbonate (Fig. 9-3), which forms mounds; many of them cover hundreds of acres and are 50 feet (15 m) or more high. The deposits collect around woody parts of plants that die and decay, leaving a highly porous mass known as tufa. Where water temperatures are high, the deposits may contain considerable silica. Most hot springs release carbon dioxide, and many release hydrogen sulfide. Sulfur-rich springs may deposit sulfates. Iron is generally present and can be tasted if in amounts greater than about one part per million. Boron is present in appreciable amounts in many of the deposits. Most warm or hot springs have an even flow, but some erupt as geysers (Fig. 9-4).

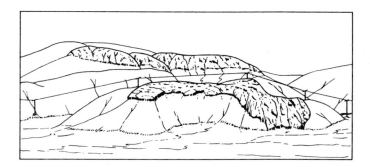

Figure 9-3. Hot-spring deposit along a fault in Death Valley, California. The spring built a mound of travertine on the hillside and extended the deposit down the bank of the arroyo (right). Evidently the travertine was deposited and the arroyo was eroded during the Pleistocene because Early Holocene (or Late Pleistocene) stone tools were found on the mound. (From Hunt and Mabey, 1966, p. A78)

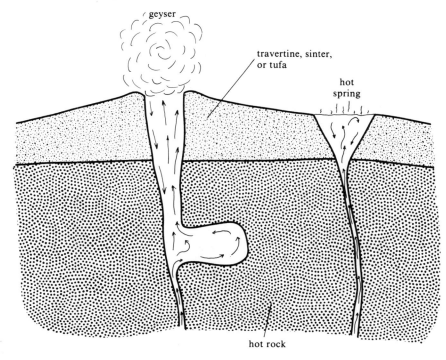

Figure 9-4. Contrast in structure between a geyser and a hot spring. Both have their sources in hot rock, but a geyser has a reservoir that explodes as steam faster than the cooled water can refill it. A hot spring is supplied at a more even rate.

The upper temperature limit for animal life in hot springs is about 122° F (50° C). Plants, however, survive higher temperatures, and the kinds depend on the temperature. Green algae, for example, flourish in water having a temperature range of about 120° to 140° F (49° to 60° C); orange and red algae flourish in water with temperatures of about 140° to 160° F (60° to 71° C), and white algae in hotter water (Waring and others, 1965, p. 9).

Probably most of the spring deposits shown on the map are of Pleistocene age, some are approximately dated archaeologically. Older springs no doubt were numerous during the extensive Late Tertiary volcanism, but most of these older deposits have been destroyed by erosion or are buried under younger deposits.

REFERENCES AND ADDITIONAL BIBLIOGRAPHY

Baldwin, B., and W. R. Muehlberger, 1959, Geologic studies of Union County, New Mexico, *New Mexico Bur. Mines and Min. Resources Bull. 63*, 171p.

Bowie, A., 1914, 1915, The burning of coal beds in place, *Am. Inst. Mining Eng. Bull.* **86:**195-204 (1914); Tr. 48; p. 180-193 (1915).

Gilbert, G. K., 1890, Lake Bonneville, U.S. Geol. Survey Monograph I, 438 p.

Hamblin, W. K., 1974, Late Cenozoic volcanism in the western Grand Canyon, in *Geology of Grand Canyon*, W. J. Breed and E. C. Roat, eds., Museum of Northern Arizona, Flagstaff, 185p.

Hamblin, W. K., and M. G. Best, 1970, The Western Grand Canyon District, *Utah Geol. Soc. Guidebook to the Geology of Utah, No. 23.*

Hunt, A. P., 1960, Archaeology of the Death Valley salt pan, California, *Utah Univ. Anthro. Papers No. 47*, 313p.

Hunt, C. B., and D. Mabey, 1966, Stratigraphy and Structure, Death Valley, California, *U.S. Geol. Survey Prof. Paper 494-A*, pp. A78-A79.

Jahns, R. H., A. R. McBirney, and R. V. Fisher, 1974, Igneous rocks, in *Encyclopedia Britannica*, vol. 9, pp. 201-233.

Lee, W. T., 1926, An ice cave in New Mexico, *Geog. Rev.* **16**:55-59.

McKee, B., 1972, *Cascadia, the geologic evolution of the Pacific Northwest:* McGraw-Hill, New York. Especially Chapters 15, 16, and 17.

Oakeshot, G. B., 1971, *California's Changing Landscape*, McGraw-Hill, New York. Especially Chapter 15.

Rogers, G. S., 1917, Baked Shale and Slag Formed by the Burning of Coal Beds, *U.S. Geol. Survey Prof. Paper 108*, pp. 1-10.

Stearns, N. D., H. T. Stearns, and G. A. Waring, 1937, Thermal Springs in the United States, *U.S. Geol. Survey Water-Supply Paper 679-B*, pp. 59-206.

Waring, G. A., revised by R. R. Blankenship and R. Bentall, 1965, Thermal Springs of the United States and Other Countries of the World—A Summary, *U.S. Geol. Survey Prof. Paper 492*, 383p.

Waters, A. C., 1955, Volcanic Rocks and the Tectonic Cycle, in *Crust of the Earth*, A. Poldervaart, ed., Geol. Soc. America Spec. Paper 62, pp. 703-722.

White, D. E., 1968, Hydrology, Activity, and Heat Flow of the Steamboat Springs Thermal System, Washoe County, Nevada, *U.S. Geol. Survey Prof. Paper 458-C*, 109p.

INDEX

ABOUT THE AUTHOR

CHARLES B. HUNT began studying surficial deposits during the 1930s when he mapped the Henry Mountains area, Utah. G. K. Gilbert's classic work there (1875-1876) described "hills of planation" (pediments), erosion of badlands and divides, as well as the intrusive structures. Later, Hunt continued following Gilbert's broad trail in working on alluvial fans, eolian deposits, and shore deposits of Lake Bonneville and other parts of the Great Basin.

During World War II, Hunt was in charge of the U.S. Geological Survey's Military Geology Unit predicting terrain conditions and water supply in the areas to be attacked. That work demonstrated the practical usefulness of surficial geology in all kinds of land use planning. It provided great impetus to the development of engineering geology.

During the 1950s, while in charge of the Survey's Regional Geology Branch, Hunt visited field parties working on surficial deposits in almost every state, and he mapped the deposits in the Denver area. During the 1960s he was professor at The Johns Hopkins University, offering courses on physiography of the United States and geology of soils, the subjects of his two textbooks.

During the 1970s, at New Mexico State University, he prepared for the New Mexico Bureau of Mines the 1/500,000 map of surficial deposits of New Mexico, the first such map of a state south of the glacial drift border. He also compiled the 1/7,500,000 map for which this book was written. Moving next to Salt Lake City he prepared, for the USGS, a 1/1,000,000 map of the surficial deposits of Utah. Finding that scale grossly inadequate he began upgrading the map to 1/500,000, a project still in progress along with a modest start towards a 1/2,500,000 map of conterminous United States.